Science and Engineering in High-Throughput Biology

including a
Theory on Parkinson's Disease

edited by

André X. C. N. Valente

Center for Neuroscience and Cell Biology, University of Coimbra

Systems Biology Group, Biocant - Biotechnology Innovation Center

Cantanhede, Portugal

ISBN 978-1-257-11175-6

Contents

Contributors

Lino Ferreira, Ph.D.

Biomaterials and Stem Cell Research Group, Biocant - Biotechnology Innovation Center, Cantanhede, Portugal

Center for Neuroscience and Cell Biology, University of Coimbra, Cantanhede, Portugal

Stephen S. Fong, Ph.D.

Dept. of Chemical and Life Sciences and Center for the Study of Biological Complexity, Virginia Commonwealth University, Richmond, Virginia, USA

Tiago F. Outeiro, Ph.D.

Cell and Molecular Neuroscience Unit, Instituto de Medicina Molecular, Lisboa, Portugal

Instituto de Fisiologia, Faculdade de Medicina, Universidade de Lisboa, Lisboa, Portugal

Jorge A. B. Sousa, M.S.

Systems Biology Group, Biocant - Biotechnology Innovation Center, Cantanhede, Portugal

A. X. C. N. Valente, Ph.D.

Systems Biology Group, Biocant - Biotechnology Innovation Center, Cantanhede, Portugal

Center for Neuroscience and Cell Biology, University of Coimbra, Cantanhede, Portugal

Center for the Study of Biological Complexity, Virginia Commonwealth University, Richmond, Virginia, USA

Preface

Technological advances are changing the panorama in the Biological Sciences. Continued improvement in instrumentation is allowing new types of observations and manipulations to be made at the cellular and molecular level. As importantly, the assay, the observation, the construct, all elements that traditionally required painstaking individual, manual caretaking, can now increasingly be handled in automated fashion. In the experimental front, Biology is thus shifting from its data-poor origins to a quantitative data, high-throughput regime.

A pressing matter is how to develop biological theory, hopefully now with greater predictive power, in this new data-rich regime. Grappling with impossibly large amounts of information, a common route has been to leave the analysis strictly to mathematics. This has led to the rise of the hypothesis-free approach. We believe it is a poor choice. Progress, even in the high-throughput regime, must continue to rely on biological intellectual insight. Now, substantial quantities of data, by necessity, require mathematical handling. Therefore, an ability to mathematically express biological knowledge and hypotheses is crucial for successfully exploiting high-throughput data. The article "Prediction in the Hypothesis-Rich Regime" proposes Rationality Classes as a transparent and effective approach to translate between biological thought and mathematics. Section 1 discusses the fundamental problem of prediction in the new biological high-throughput regime and introduces the Rationality Classes framework. The original "Prediction in the Hypothesis-Rich Regime" article is presented in Section 2.

Medicine stands to greatly benefit from fulfillment of the scientific and engineering potential of high-throughput biology. As a practical implementation of the Rationality Classes approach, we analyzed genome-wide expression data from Parkinson's Disease patients. The ensuing finding that a broad number of functional processes known to be implicated in Parkinson's Disease appear to be also disrupted in the blood cells of Parkinson's patients served as a starting point for us to propose a wider re-evaluation of the character and origin of the disease. Section 3 summarizes the theory on Parkinson's Disease we are proposing. Section 4 presents the article "A Stem-cell Ageing Hypothesis on the Origin of Parkinson's Disease" where our view was originally presented.

Besides its impact in Pure Science, technology is also enabling implementation of complex engineering solutions in the biological setting. A framework for engineering design in this context is therefore essential. Biological knowledge is vital for the design process. But, given the inherently limited predictive ability of science in the biological setting, we argue biological knowledge will often be most fruitfully applied to rationally optimizing a high-throughput search process in the space of design possibilities. We call this approach High-Throughput Biologically Optimized Search Engineering (HT-BOSE) and argue for its relevance in Section 5. Section 6 contains the original article "High-Throughput Biologically Optimized Search Engineering Approach to Synthetic Biology". The article presents a HT-BOSE foundation for Synthetic Biology, the emerging field of engineered complex biological systems.

A. X. C. N. Valente

1

Science in the Biological High-Throughput Regime

The driver of scientific progress in Biology has been overwhelmingly the experimental component. Historically, more often than not, the relevant theoretical implications of new experimental observations have been readily apparent.

Ongoing developments enabling the collection of a diversity of biological data in a high-throughput fashion are creating an explosion in the amount of quantitative data available in the Life Sciences. Notably, this time the repercussions are not so obvious. Expectations are that the new data will significantly increase understanding and the ability to predict events in biological systems. But this is not a certain outcome. In reality, it is not even clear what is the best approach to attempt conversion of this raw data into actual science: laws that synthesize knowledge and give us predictive power.

The elementary answer to the above is that doing science in the high-throughput biological context is no different from doing science in a more traditional setting. You have to draw upon your knowledge, think and put forward insightful hypotheses. There are no sweat-free automated methods for converting raw data into beautifully synthesizing laws. This may be biologically intuitively obvious. Significantly, it is also a correct mathematical statement: An absolutely hypothesis-free approach to data analysis is mathematically assured to produce absolutely no results. In other words, regardless of computational power, a mathematical approach devoid of biological content is doomed to fail. This observation amounts to an expression of the limitations and the frail foundation of Science, whose modern discussion remits us to the works of David Hume (1711-1776).

Now, there is a defining, critical characteristic shared by most scientific problems in the biological high-throughput data context. It is the presence of a large number of a priori plausible alternative explanations for the observed phenomena. This attribute defines what I call the "hypothesis-rich" regime.

The article "Prediction in the Hypothesis-Rich Regime" puts forward an approach for doing science in this regime, so prevalent in the Life Sciences today. It proposes tackling a problem by thinking in terms of Rationality Classes of potential laws. It presents this as the clearest way of leveraging biological knowledge, hypotheses and insights in the hypothesis-rich regime. It further shows how to mathematically enable this approach.

The article is written in an abstract, formal manner. This may make its essential ideas - and their implications - a little more difficult to come across. We will thus now discuss some of them, or at least my personal interpretation of them, in a more colloquial fashion.

We shall do this in the context of a motivating example. Let us consider a disease diagnostic problem. Based on gene expression data, we wish to classify individuals into one of two classes, healthy or diseased (Figure 1). In this context, a law is any division of the gene expression space in two regions, one predicting healthiness, the other predicting disease (Figure 2). In general, a law is any rule for making predictions based on some initial information.

We associate with every law two values, a True Quality (TQ) and an Observed Quality (OQ). The TQ measures how good a law is at predicting outcomes over all cases of interest. We define as the final objective to obtain a law with a high TQ. A TQ metric that matches our goals for the problem in question should be selected. In our example problem, a reasonable choice for the TQ of a law would be its percentage of correct predictions over the population of interest. Now, the TQs of laws are unavailable to us. In our problem, they would require us to know, for every person in the population of interest, its gene expression and whether it was healthy or diseased. Selecting the highest TQ law would be trivial in that case. Instead, all we have available is a limited number of individuals, for whom both gene expression data and the health/disease state were recorded. Based on this restricted information, we can assign to a law an OQ. This is the counterpart to its TQ, except it is based solely on the available observations and is thus computable. The OQ metric serves as our best estimate, based exclusively on the observed data, of the desirable but

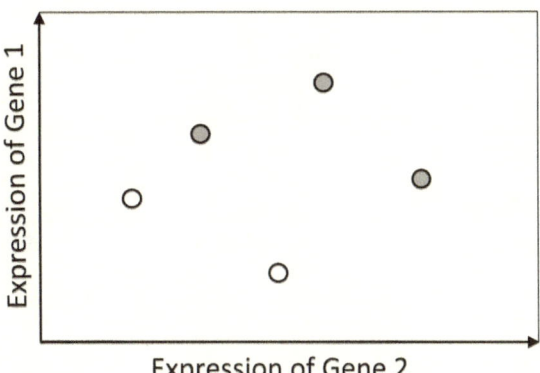

Figure 1. A typical problem in the high-throughput biological regime is the search for a law that, based on genome-wide measurement of gene expression levels in a tissue or fluid sample from an individual, predicts whether the individual has the disease of interest. Joint gene expression data and the actual health/disease state are available for a limited number of individuals. Above, only 2 of the 20 000 plus dimensions of gene expression space are shown.

Examples of Laws

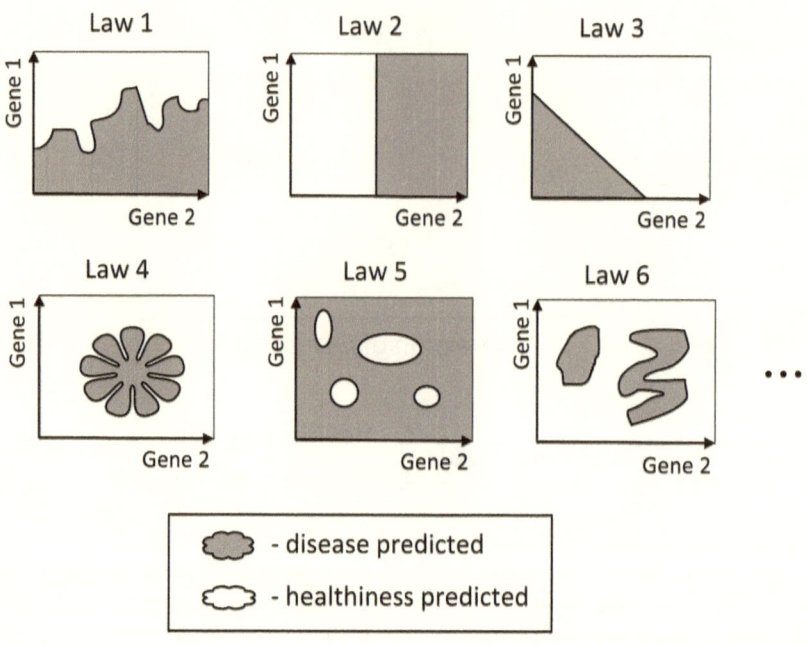

Figure 2. In the gene-expression based disease diagnosis problem, a law is a split of the 20 000 plus dimensional gene expression space into two regions, one assigned to health, the other assigned to disease.

unobtainable TQ of a law. In our example, it would be acceptable to define the OQ of a law as its percentage of correct predictions over the recorded cases. Unfortunately, OQ is bound to be stochastically off from TQ. First, because it is based on a limited sampling of the cases of interest. Second, because there are invariably experimental outcome measurement errors. In our example, some of the individuals recorded as healthy may in fact be diseased. Note that in the article "Prediction in the Hypothesis-Rich Regime", instead of TQ and OQ, the formalization is in terms of True Distance and Observed Distance to an ideal, perfect law. These are respectively akin to TQ and OQ, just on a reversed scale, that is, low True Distance corresponds to high TQ and similarly low Observed Distance corresponds to high OQ.

Let us consider a hypothesis-free approach to finding a good law. This assumes placing all potential laws on an equal footing. In our example, the set of all possible laws is the set of all possible divisions of the gene expression space in two regions, assigned respectively to health and disease. Call it the Full Set. By construction, the comparatively few high TQ best laws are included in the Full Set. But the Full Set also contains the massive number of very bad low TQ laws. Combined with the stochastic error in the OQs estimates of TQs, this results in most of the high OQ laws being in fact low TQ laws. It becomes impossible to select a high TQ law via OQ, as statistically the few high TQ laws are OQ-wise drowned amongst low TQ laws (Figure 3, left panel). Note how this failure is not due to a lack of computational power to analyze the data, but rather of a more fundamental nature.

The reason analyses deemed "hypothesis-free" produce results is that in fact they are not hypothesis-free. In our example problem, it would be common to analyze the gene expression data by "hypothesis-free" assessing the potential of every individual gene as a biomarker. This is akin to considering only splits of the gene expression space defined by a critical expression-level of an individual gene. This is an extremely constrained subset of the Full Set of possible splits of gene expression space. Restriction to this limited set of laws is rooted on biological knowledge. Namely, on knowledge that genes are fundamental functional units in organisms and therefore that it would make particular biological sense for a critical level of expression of a gene to be the determinant of health vs. disease. In contrast, the disease and health regions for some of the laws shown in Figure 2 do not make biological sense.

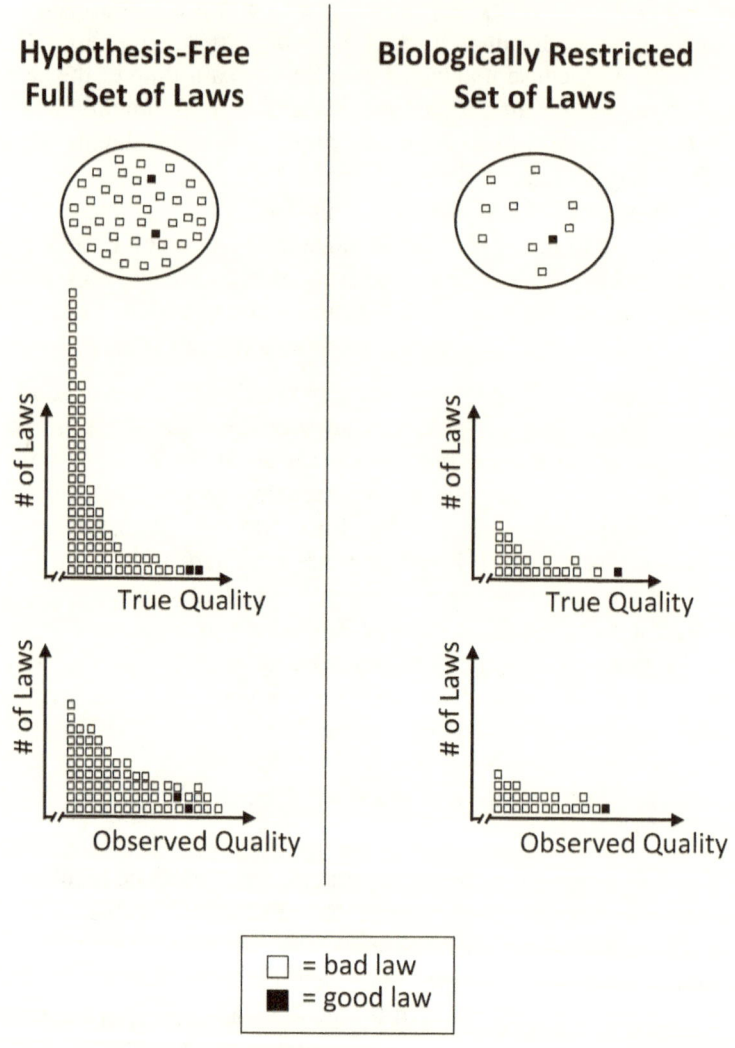

Figure 3. For illustrative purposes, the higher quality laws are shaded in black. Under a hypothesis-free approach, all possible laws are given equal consideration. The overwhelming number of low True Quality laws, combined with stochasticity, results in the high True Quality laws being Observed Quality-wise drowned amidst lower True Quality laws. Alternatively, biological knowledge can be used to restrict the set of laws under consideration. Above, the favorable True Quality Distribution and Correlation Structure of this biologically restricted set permits the highest Observed Quality law to still correspond to a high True Quality law.

In sum, biological knowledge is used to a priori restrict consideration to a rational set of laws. Note this is always the case in successful analyses, it is just sometimes not explicitly stated, or directly apparent. The restriction is mathematically successful if the *TQ Distribution and Correlation Structure* (TQD) of the set of rational laws is favorable enough for a high OQ law to be statistically likely indeed a high TQ law (Figure 3, right panel). Note in the article "Prediction in the Hypothesis-Rich Regime", in keeping with the formalization in terms of True Distance and Observed Distance, TQD is replaced by TDD (True Distance Distribution and Correlation Structure).

In general, different biological arguments lead to different sets of rational laws. We call these sets Rationality Classes (RCs). In our example problem, the commonly used and alluded to above RC, containing all laws based on a critical level of expression of a single gene, is one biologically sensible choice. But, based on a belief that the disease is a metabolic malfunction, an analogous RC except further restricted to metabolism-related genes would be reasonable too. A yet alternative theory about the disease, may argue that crucial to the pathology is an unbalance in the expression level of pro- and anti-apoptotic genes. Then, a RC of laws based on ratios between pro- and anti-apoptotic genes would seem appropriate. Suppose based on current biological knowledge, we have no preference amongst the above three rationales. A key insight is that, in such situations, it is still best to analyze the RCs separately. The distinct rationale behind each of the RCs is enough to ensure that likely the RCs have different TQDs. In a comparison, the TQD of a merger of the three classes is at most as favorable as the best of the three individual RCs TQDs. However, it could be much worse. For instance, if one of the individual RCs has a very unfavorable TQD, where many low TQ laws prevail at the highest OQs, this will effectively set the quality of the TQD of the merged class. Thus, merging of classes may result in a loss of statistical power. Conversely, an insightful biological rationale for separating a RC into two separate RCs can increase statistical power, even if we are unable to attribute a preference for one of the new RCs over the other. This added statistical power follows from the two resulting RCs possibly having rather distinct TQDs, by virtue of the biological rationale that differentiates them.

The amount and quality of the observed data can affect which TQD is more favorable. In the example problem, suppose the absolutely highest TQ law is based on a particular ratio between a pro- and an anti-apoptotic gene. Let the other laws in the ratios based RC have low TQs. In contrast, assume in the metabolic genes based RC laws have mostly average TQs. Given enough observed data, the ratios based RC is the best, as it allows identification of the overall highest TQ law. But under a sparse, low quality dataset, stochastic differences between the TQ and the OQ of a law become very large. Then, some of the low TQ laws in the ratios based RC may have a higher OQ than the OQ of that highest TQ law. Under this weak dataset, the metabolic gene based RC is the best, as its highest OQ law likely still has an average TQ.

We now summarize the discussion so far. We define a TQ metric that is commensurate with our goal for the problem at hand. The ultimate objective then becomes to select a high TQ law. The TQs of laws are unavailable to us. We define an OQ metric to provide our best computable estimate of TQ based exclusively on the observed data. A RC is a set of candidate laws that, based on biological thought, share a common rationale as candidates and are thus on an equal footing, prior to leveraging of the observed data. That is, RCs are based exclusively on prior biological knowledge, OQ exclusively on the problem's observed data. It follows from the above that given a RC the best option is to select the law in the RC with the highest OQ. Whether that law is actually a high TQ law, depends on whether the TQD of the RC was favorable enough for this to occur. Success thus equates with having a biological insight that yields a RC with such a favorable TQD. Fortunately, multiple RCs, based on different biological rationales and hypotheses, can and should be proposed, in an attempt to find that favorable RC (e.g.: 291 RCs are constructed in the Parkinson's blood gene expression analysis problem of a later section). In any given problem, the major effort should thus be in building appropriate RCs. Notably, an insight that allows splitting a RC into two (or more) RCs can be key, regardless of ability to assign differential preferences to the resulting RCs. This effectively lowers the bar on what is demanded of biological knowledge to make a positive impact and thus should be kept in mind when thinking biologically on how to construct RCs.

The presented viewpoint also makes transparent the mathematical limits on finding a good law. Consider a RC where 505 of its laws possess an equal top OQ score (in our example, say the 505 laws do identically well at predicting the disease state of the observed individuals). Suppose in reality only 5 of these laws have a high TQ, while the other 500 have a low TQ, i.e., their high OQ was a result of chance. By symmetry of the situation, it is impossible for a purely mathematical or statistical method to discriminate between these laws.

Note how this holds irrespective of computational power for analyzing the data. To pinpoint one of the high TQ laws, further biological insight is required. For instance, an insight that leads to splitting of the RC into two new RCs with distinct TQDs, in this case with one of the new RCs containing at least one of the aforementioned high TQ laws but none of the other 500 high OQ, low TQ laws. Awareness of such limitations of mathematical and computational power can save pointless efforts.

We now describe an elementary mathematical implementation of the Rationality Classes approach (Figure 4). In the example problem, we split the individuals that constitute the observed data into three separate datasets, 1 thru 3. In a first stage, we use dataset 1 to assign OQs to the laws in the RCs. From each of the constructed RCs (by chance, also three in our example), we select the law (or a law) with the highest OQ. In a second stage, those three selected laws are assigned new OQs based on dataset 2. The highest stage 2 OQ law is our final selected law. In a third stage, using dataset 3, we compute and report a final OQ estimate of the TQ of our selected law. The first stage provides the best candidate law from each RC. The second stage permits weeding out a high OQ (at stage 1) but in reality low TQ law put forward by a RC with an unfavorable TQD. The third stage provides for the final selected law an unbiased OQ assessment of its TQ. Note how the biological biases introduced in the law search do not preclude an unbiased assessment at the end. We are thus presenting a *biased search, unbiased assessment* approach.

Collectively, the laws selected for the second stage effectively form a new 2nd stage RC from where the dataset 2 based highest OQ law will in turn be chosen. This 2nd stage RC has its own TQD. The discussed dynamics involving OQs, TQs and TQD, are therefore fully applicable again. Naturally, with a rather limited number of RCs, we expect the TQD of this 2nd stage RC to be favorable, i.e., we expect all 2nd stage OQs to be accurate enough to make possible the weeding out of low TQ laws that by chance had a high OQ at stage 1. On the other hand, this also shows that careless formation of too many RCs can result in an unfavorable 2nd stage TQD that precludes selection of a high TQ final law.

Creation of an increasing number of arbitrary RCs at the 1st stage eventually produces by chance a RC with a favorable TQD. Given that this requires no biological insight, has the principle that hypothesis-free approaches cannot work been circumvented? No, because the growing number of RCs also leads the TQD of the 2nd stage RC to gradually

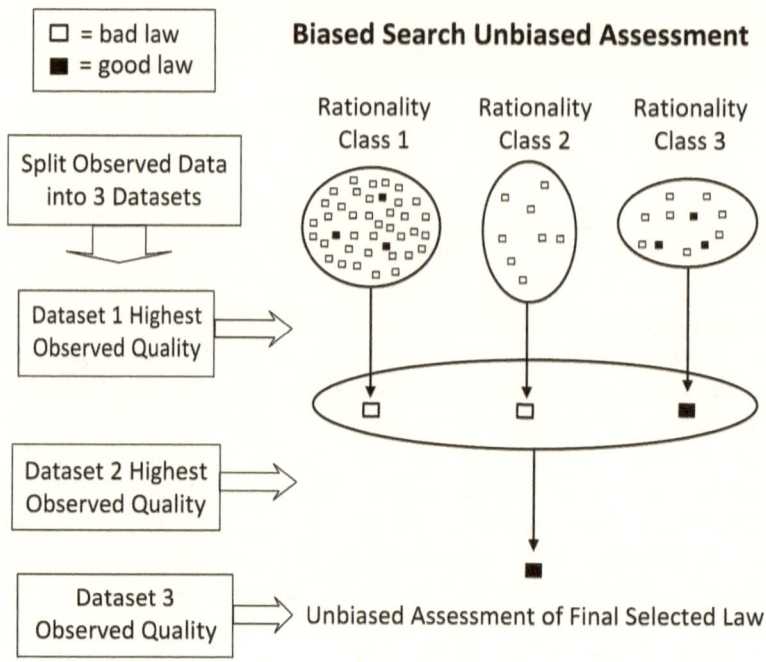

Figure 4. An elementary mathematical implementation of the Rationality Classes (RCs) approach is shown. Distinct biological rationales and hypotheses permit formation of separate RCs, even if no differential preference can be attached to these. Dataset 1 is used to assign Observed Quality (OQ) scores to the laws and to select the highest OQ law from each RC. These selected laws are then assigned new OQ scores based on Dataset 2. The new OQ highest scoring law becomes the final selected law. Dataset 3 is used to perform an OQ assessment of the True Quality of the final selected law that is not biased by evaluation of multiple laws. Above, only the Rationality Class 3 True Quality Distribution and Correlation Structure was favorable enough for its highest OQ law to correspond to a good (i.e., high True Quality) law. Given that the OQ of this good law is lower than the OQs of many of the laws in Rationality Class 1, had the three RCs been analyzed as a single class, this good law would not have been detected.

become unfavorable, thus precluding, now at this 2nd stage, the selection of a high TQ law. The principle that hypothesis-free methods cannot work is therefore not violated. Everything is logically consistent.

The faithful replication of the setup at the 2nd stage points to framing the approach in a self-similar manner (Figure 5). Just as laws are grouped into RCs, these RCs may in turn be grouped into higher-level RCs. If backed by a biological rationale, these higher-level RCs may have distinct TQDs. Therefore keeping them separate can be advantageous, for the same reasons expounded for not merging basic RCs. The process can be repeated with the higher-level RCs being in turn grouped themselves into further higher-level RCs and so forth. Biological knowledge and hypotheses are then mathematically represented in a hierarchical tree structure of laws, RCs and higher-level RCs.

The price for creating a hierarchy with multiple RC levels is the required splitting of the observed data into independent datasets, one for each of the hierarchical levels. Statistical power is decreased by the smaller datasets yielding larger stochastic errors in their OQ estimates of TQ. In this regard, the described elementary division of the observed data into disjoint datasets is far from optimal. In practice, the compact recursive implementation presented in the article "Prediction in the Hypothesis-Rich Regime" (see article figure 4) better optimizes the use of the available data, while making apparent the self-similar nature of the framework. That said, in any implementation there will always be an additional cost for each extra hierarchical level. Note that the framework requires at least three hierarchical levels, if the final OQ estimate of the select law is included. The proposition, of course, is that in general this loss is more than compensated by the ability to keep apart RCs with distinct TQDs.

The article "Prediction in the Hypothesis-Rich Regime" lays the groundwork of a Rationality Classes based approach to Science in the biological high-throughput regime. It is a biased search, unbiased assessment approach. The concept of RCs permits a clean translation from biological thought to mathematics. This is crucial, as ultimately success hinges on mathematically leveraging biological knowledge and hypotheses. Further, taking this biologically biased approach may benefit the reverse translation at the end, that is, the mathematical results become easier to interpret biologically. Can the mathematical TQ or OQ measures be improved? Is increased computational power

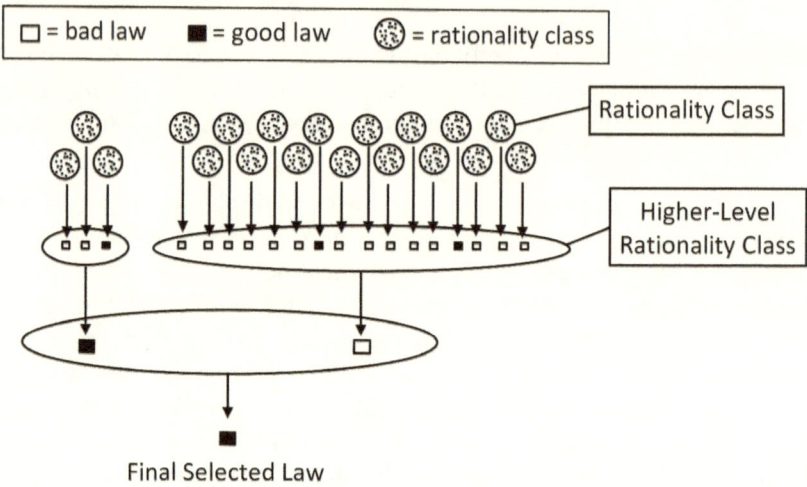

Multi-level Rationality Classes Structure

Figure 5. Biological knowledge may be used to further group Rationality Classes (RCs) into higher-level Rationality Classes. Above, the higher-level RC on the right has an unfavorable True Quality Distribution and Correlation Structure. Separation of the two higher-level RCs still permitted the final selected law to be a good one.

necessary or irrelevant? Is there further biological knowledge to leverage? The Rationality Classes approach sound construction around the fundamental concepts of TQ, OQ, TQD and RC results in a transparent analysis that facilitates such essential troubleshooting.

This discussion and the article that follows introduce only the overarching principles for this approach. Certainly, matters are raised that are only barely touched upon. Furthermore, each concrete problem will require its own tailor-made practical implementation.

A. X. C. N. Valente

2

Article

Prediction in the Hypothesis-Rich Regime

(arXiv: March 18th, 2010)

Prediction in the Hypothesis-Rich Regime

André X. C. N. Valente[1,2]

[1]Center for Neuroscience and Cell Biology, University of Coimbra, Portugal

[2] Systems Biology Group, Biocant, Cantanhede, Portugal

We describe the fundamental difference between the nature of problems in traditional physics and that of many problems arising today in systems biology and other complex settings. The difference hinges on the much larger number of a priori plausible alternative laws for explaining the phenomena at hand in the latter case. An approach and a mathematical framework for prediction in this hypothesis-rich regime are introduced.

Introduction

Science can be seen as an attempt to find good laws, rules that given some observed input information, are able to predict accurately a resulting, yet unobserved, outcome of interest [1]. Experimental observations guide this search, by allowing proposed laws to be matched against a body of data where both input information and respective outcome are available. Even in traditional physics problems, it is important to realize that generally there exist a multiplicity of laws that replicate well the available experimental data. For example, planetary motion is very accurately predicted by Newton's inverse square gravitational law. But consider an alternative law stating that this is true up to the year 2100 AD, beyond which the force decreases cubically, rather than quadratically with the distance. Based solely on matching available planetary motion experimental data, this other law is exactly as good as Newton's law. Similarly, as shown by both Ptolemy and Copernicus, it is possible to construct an epicycle based theory that correctly predicts available planetary motion data [2]. The reason we dismiss these alternatives is that, unlike Newton's law, we consider them irrational.

The objective is therefore to find a law that i) is rational and ii) matches available data. Contributors to determining the rationality of a law include both a) information beyond the data in item ii) above, from observations not directly related to the predictions of the law and b) fundamental beliefs that we hold to be true. For instance, rationality requires assuming properties such as the underlying homogeneiety of space and time. Minimal good-behavior of any involved mathematical functions is generally expected in a rational law. Laws based on conservation of some mathematically defined quantity are often considered particularly rational too. As a final example, the simplicity of a law (relative to the problem at hand) is usually equated with its rationality, i.e., the Occam's Razor principle.

The nice property of most typical problems in physics is that, once an arguably rational law that matches the observed data is found, we can be fairly confident that it will predict equally well the yet unobserved Cases. This is because in these problems the rationality constraint we are imposing is restrictive enough that it is very unlikely that an arguably rational law would match available experimental data just by chance. However, in contrast with the above situation, the typical systems biology [3] problem involves many a priori plausibly relevant variables and generates many more rational alternative hypothesis.

Let us introduce one possible formalization of these issues (Fig. 1). Define a finite measure [4] on the Space of Cases of Interest and a metric on the Space of Outcomes. Together, these induce a metric on the Space of Laws (a space of mappings from the Space of Cases of Interest to the Space of Outcomes). Call it the True Metric [5]. Let the idealized Perfect Law map every Case to exactly its correct outcome. Now associate with each law a *True Distance*, the distance between that law and the Perfect Law under the True Metric. The True Distance of a law gives an averaged distance between its predicted outcomes and the correct outcomes. The True Distance of a law is unknown to us. Our ultimate objective is to find a law with a low True Distance (with the practical restriction of mapping all Cases with identical observed input information to the same Outcome). Define an *Observed Distance*, analogously to the True Distance, except in that: i) It is based solely on the Cases with available experimental data [6] and ii) the term of comparison is not the correct outcomes of the Perfect Law, but rather those experimentally measured outcomes. The Observed Distance of a law gives an averaged distance between predicted and observed outcomes for the set of Cases with experimental data. Observed Distance is defined to provide the best possible *computable* estimate of True Distance, based *exclusively* on the available experimental data (i.e., rationality considerations aside). Each specific scientific problem may require its more particular, tailor-made, Observed Distance definition [7]. For our purposes, only assessing the relative True Distances of laws will prove relevant.

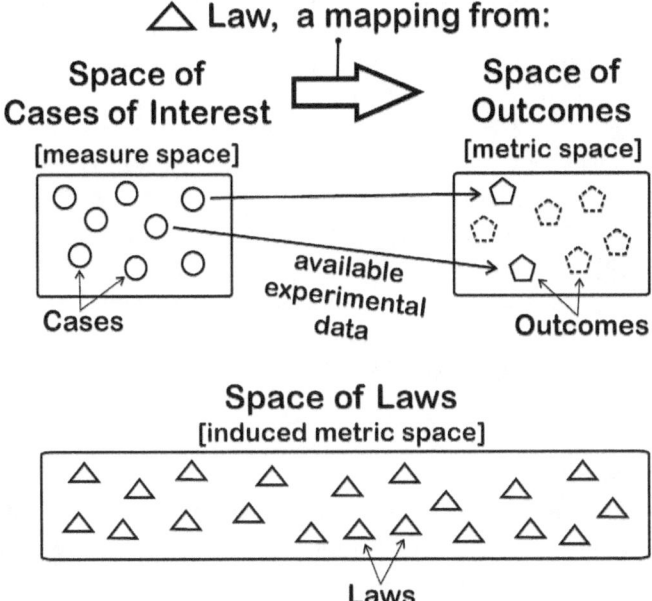

Figure 1. A formalization of the scientific problem. The Space of Cases of Interest is assumed fixed a priori. Each Case has associated with it a correct outcome and some experimentally observed input information. Measurement noise is handled by considering as distinct Cases even Cases that differ only in the observed input information. The goal is to find a law that can accurately predict a Case outcome, given the Case input information (i.e., subject to mapping all Cases with the same observed input information to the same outcome). To help, there is a body of (noisy) data on the outcomes of some Cases.

The body of available experimental data is the realization available to us of its corresponding random variable. The computed Observed Distance of a law is therefore also the available realization of its corresponding random variable. Since usually the experimentally observed data covers but a small subset of the Space of Cases of Interest and, additionally, the observed outcome measurements can be noisy, Observed Distances are going to be off from the respective True Distances. Therefore, if the set of laws we consider rational is too large, in particular if it contains too many laws with a high True Distance, then, statistically, likely there will be high True Distance laws with Observed Distances as low or lower than the Observed Distances of the low True Distance laws we would like to select out (Fig. 2). This difficulty is the so called over-fitting, or under-determination of a problem [8]. Unlike in traditional physics, the nature of problems arising in systems biology and other complex settings [9–11] is such that at least the possibility of this occurring tends to be present. We next introduce an approach for doing science in this fundamentally different hypothesis-rich regime.

Rationality Classes

Let the final aim be to select a single law, with a True Distance as low as possible. Computational time considerations in the search and testing of candidate laws are disregarded. We i) define a Set of Candidate Laws based on rationality considerations (defined exclusive of the information in the experimental data used in the Observed Distance computation) and then ii) select the law from this set with the lowest Observed Distance. The likely quality (True Distance) of the selected law is determined by the *True Distance Distribution and Correlation Structure* (henceforth shorthanded TDD) of the Set of Candidate Laws [12]. Namely, it is determined by what True Distance values are statistically likely to be prevalent at the lowest Observed Distance (Fig. 3, Basic Method). The search for a good law therefore can be equated with the search for a set of laws with a favorable TDD. The set of all possible laws will usually have a terrible TDD that leads to the selection of a bad law. Defining a rational Set of Candidate Laws is an attempt to obtain a more favorable TDD.

Typically a variety of presumed equally rational arguments can be put forward to generate a rational set of laws. Call the set of laws generated by each such argument a Rationality Class. As an example, consider prediction of tumor growth/evolution [13–15]. Suppose a particular set of biological considerations about cancer growth leads to a model with seven partial differential equations and five undetermined parameters.

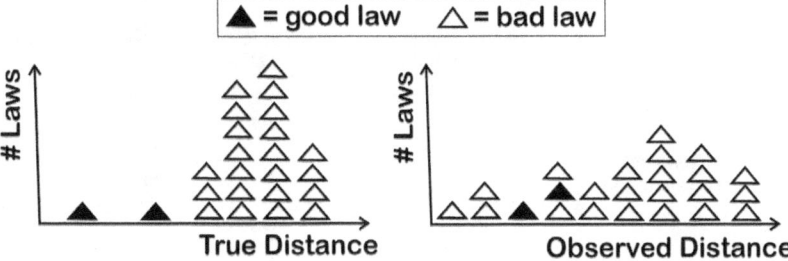

Figure 2. The presence of too many (bad) laws with a high True Distance (left chart) will, Observed Distance wise, occlude the low True Distance (good) laws (right chart).

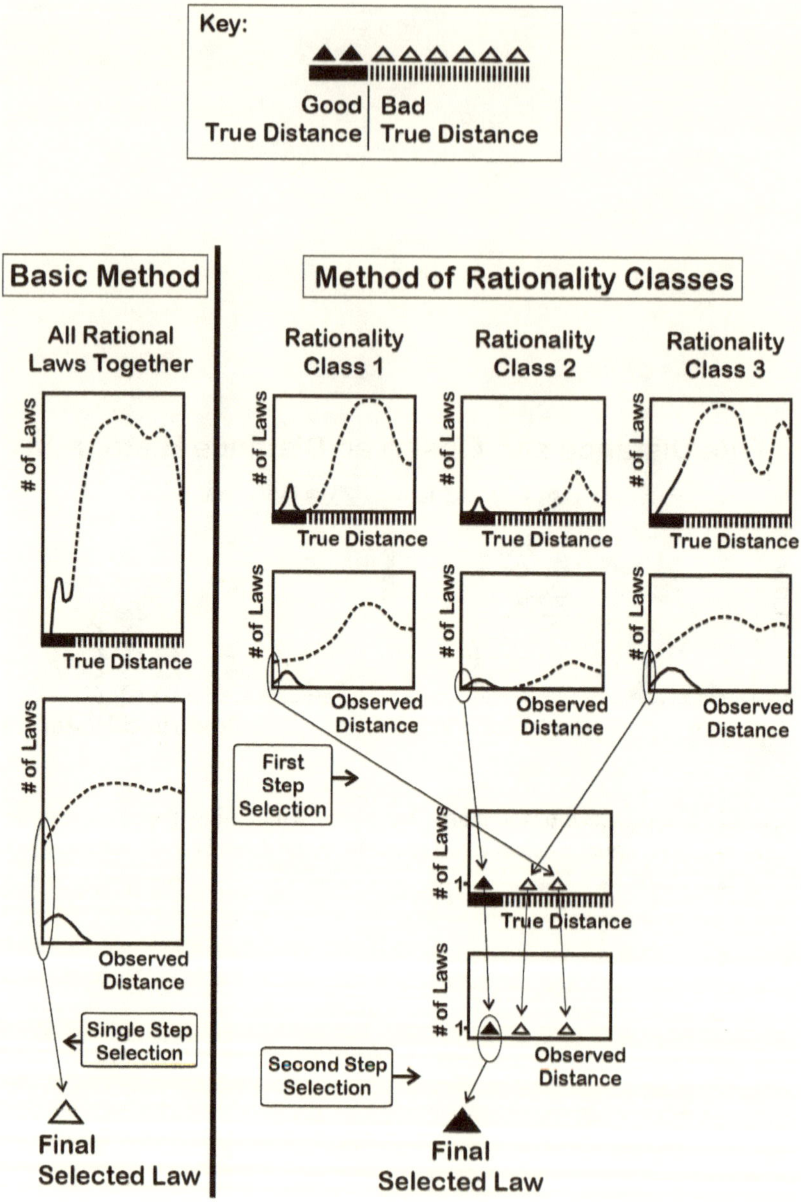

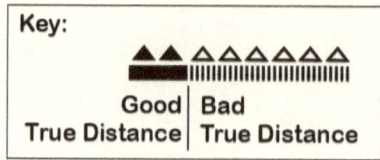

Key:

Good | Bad
True Distance | True Distance

Basic Method

All Rational Laws Together

of Laws

True Distance

# of Laws

Observed Distance

Single Step Selection

Final Selected Law

Method of Rationality Classes

Rationality Class 1

of Laws

True Distance

# of Laws

Observed Distance

Rationality Class 2

of Laws

True Distance

# of Laws

Observed Distance

Rationality Class 3

of Laws

True Distance

# of Laws

Observed Distance

First Step Selection

# of Laws

1 —

True Distance

# of Laws

1 —

Observed Distance

Second Step Selection

Final Selected Law

Figure 3 (Opposite page). Two alternative methods for selecting a final law. For distinction, solid lines denote good True Distances, dashed lines denote bad True Distances. Similarly, a solid triangle represents a law with a good True Distance, an open triangle represents a law with a bad True Distance. *Basic Method:* a rational Set of Candidate Laws is defined. Its True Distance Distribution and Correlation Structure is unknown to us (# of Laws vs. True Distance histogram; note that the correlation structure is not shown). The Observed Distances of these laws, generated by the available experimental data, are known. The # of Laws vs. Observed Distance histogram shows a typical Observed Distance distribution for the good True Distance laws and similarly for the bad True Distance laws. The final law is selected from those with the lowest Observed Distance. In this example, a bad True Distance final law was selected, as most of the laws at the lowest Observed Distance had a bad True Distance. *Method of Rationality Classes:* The same rational Set of Candidate Laws this time is divided into 3 Rationality Classes, each with its own True Distance Distribution and Correlation Structure. The available experimental data is divided into two test sets. Based on the first test set, a lowest scoring Observed Distance law is selected from each of the three classes. Based on the second test set, new Observed Distances are generated for these three laws, the one with the lowest one being selected as the final law. In this example, the existence of a favorable True Distance Distribution and Correlation Structure in Rationality Class 2, together with the two-level hierarchical procedure, allowed a good True Distance law to be sifted as the final selected law.

This model defines a Rationality Class, each set of values for its parameters giving rise to a law in that class. Experimental data taken on the evolution of a number of actual tumors could then be used to chose the best values for these five parameters. This is the process of selecting the law with the lowest Observed Distance in the class. But perhaps additional biological arguments could be raised, that if true would either settle some of these parameters or constrain their possible range. The model under this restricted parameter space would define a second Rationality Class. On the other hand, arguably, cancer growth prediction is really best if modeled at a more detailed scale. This could lead to a set of thirty partial differential equations with twenty undetermined parameters. This would constitute yet another distinct Rationality Class.

Most importantly, even though we are unable to determine a priori which of the Rationality Classes is the best, we know their distinct origin gives them a fair chance of having significantly different TDDs. A key insight is that, in such cases, merging these classes into a single Set of Candidate Laws and selecting the law with the lowest Observed Distance is often not ideal, as that law may actually not be coming from the class with the most favorable TDD.

We therefore propose a two-level hierarchical process for selecting the single final law (Fig. 3, Method of Rationality Classes). This process entails the division of the available experimental data into two separate test sets. First, from each Rationality Class, the law with the lowest Observed Distance based on the first test set is selected. The selected laws are then compared in the second step, the one amongst them with the lowest Observed Distance based on the second test set being picked as the final choice. Such a two-level process, permits effective comparison of the solutions produced by different Rationality Classes. The cost of the process is that the experimental data has to be divided into two separate smaller sets, therefore lowering in each case the accuracy of Observed Distances as estimates of True Distances. This cost can be worthwhile as some of the Rationality Classes may have substantially different TDDs. If the classes had been merged into a single Set of Candidate Laws, lower Observed Distance laws from a Rationality Class with an unfavorable TDD could occlude a higher Observed Distance but better (lower True Distance) law from a Rationality Class with a more favorable TDD. Note how the goal has moved to one of identifying a collection of Rationality Classes worth testing (hoping that the collection will contain a good class, as it in turn will yield a good final law, picked in the second step of the above process).

Consider a Rationality Class that cannot be broken down via some argument into distinct Rationality Classes. Laws in this class that further have the same Observed Distance are completely indistinguishable. In such a class, by definition, it is impossible to do better than to select the law with the lowest Observed Distance. In particular, no random breaking of the class into separate classes can overcome this fundamental limitation. Also, it is always best to test all the laws in a given Rationality Class. Not doing so is tantamount to not selecting the lowest Observed Distance law in that class.

In principle, it is possible to keep creating Rationality Classes, or to keep dividing a Rationality Class into more and more separate classes via gradually subtler arguments. Recall that each of these classes sends its selected law to the second level of the process. Together, these selected laws form their own class at the second level, with its own TDD, from where the final law is selected. The discussion pertaining to the TDD of a class is equally valid for this new class at the second level. Therefore, in particular, a non-judicious, too liberal class creation at the lower level could result in an unfavorable TDD at the second level.

Recursive Framework

We now briefly introduce a recursive framework (Fig. 4) that optimizes some aspects of this new approach to prediction based on Rationality Classes. First, note that it may be profitable to extend the concept of Rationality Classes to higher-level grouping of Rationality Classes and so forth, as shown in the Hierarchical Tree Structure of Figure 4. This is so as, analogously to what happens with basic Rationality Classes, the Higher-Level Rationality Classes (at some given higher-level) may have distinct TDDs and therefore considering them as a single Class may be detrimental. Second, the Figure 4 recursive framework helps extract more out of the set of Cases with available outcome data. It does so, i) via the repeated splitting of T into different (T*, U*) combinations when determining the best child node to select at any given generic node x and ii) via finally passing to that selected child node (Best of the Child Nodes recursive call) the full set T originally received by the parent generic node x. Note how Cases in the input set U of unknown Cases to a node x never become cases in the input Set T of available Cases to a downstream node, called within

Hierarchical Tree Structure

(more levels possible)

Root Node
(first call)

Higher-level
Rationality Class

Generic
Nodes

Rationality
Class

Law

Leaf
Nodes

Generic Node

Set T of
Available
Cases

Split T:

T*

U*

Child
Node 1

Child
Node 2

Child
Node 3

etc.

Compute
Observed
Distances
based on
the U*s to
determine
the Best
of the
Child
Nodes

Done

Best
of the
Child
Nodes

Predictions
for Set U

Set U of
Unknown
Cases

Repeat w/ different splits (T*, U*)

Leaf Node

Set U of
Unknown Cases

Law

Predictions
for Set U

Figure 4 (Opposite page). Recursive framework for prediction in the hypothesis-rich regime. *Hierarchical Tree Structure:* Laws are grouped into Rationality Classes, which in turn are grouped into Higher-Level Rationality Classes and so forth. *Generic Node:* A generic node x receives as input i) a set T of Cases with available outcome data for its use and ii) a set U of unknown Cases, where the outcome data is not provided to the node x (even if known). Node x outputs predictions for the outcomes of Cases in this set U. Inside any given generic node x, recursive calls can be made to child nodes of node x (nodes directly linked to node x in the Hierarchical Tree Structure diagram; thick border boxes inside the Generic Node diagram). Note that each node gives rise to multiple recursive threads. The repeated splitting of T into (T*, U*) can be implemented using any standard cross-validation or resampling scheme [16] (among other issues, the choice of scheme will have to factor in computational time costs, something not addressed in this article). *Leaf Node:* a recursive thread ends when a leaf node is reached. In a leaf node, a law generates predictions for the Cases in U it receives. Note that in the generic nodes that recursively call leaf nodes, T simply becomes U*, as laws have no use for the T* set. The whole process starts at the root node and ends when the root node produces predictions for its set U of unknown Cases.

a recursive thread originating at x. Thus, as proper, the process to generate outcome predictions for the set U never makes use of the set U of outcomes.

Complex Biomarkers

Although most pathological conditions are multifactorial in nature, the search for the single entity Biomarker [17] has been a dominant paradigm in medicine [18]. Methods for detecting protein interactions [19, 20], for assessing transcriptional activity [21] and for synthetic lethality screening [22] are examples from a rapidly expanding collection of techniques for acquiring biological data in a high-throughput, quantitative manner. An open question is whether integrated analysis of such data can lead to the identification of Complex Biomarkers that, based on multiple biological readings, more accurately assess the system-level status of a biological process. We view blunt integration of data into a single statistical model as a nonideal approach for finding a Complex Biomarker: It amounts to testing only one Rationality Class (Fig. 3, Basic Method). On the other hand, through appropriate biological hindsight, the multiple data sources may prove ideal to create a diversity of Rationality Classes for testing within a framework as outlined in this article.

In Valente et al. [23], we use our approach to analyze blood genome-wide expression data from 50 Parkinson's disease patients and 55 age-matched controls [24]. Specifically, a classifier law for distinguishing Parkinson's disease individuals from non-Parkinson's disease individuals based on blood expression data is constructed. An interesting result of the analysis is that it hints at differential expression in blood cells in many of the processes known or predicted to be disrupted in Parkinson's disease. Starting from this observation, a hypothesis on Parkinson's disease originating in a hematopoietic stem cell differentiation process expression program defect is proposed.

Summary

In this article we put forward that problems in systems biology and other complex settings often fall in the hypothesis-rich regime. We argue that doing science in this regime requires a fundamentally distinct approach, which is proposed. Namely, the concept of Rationality Classes is introduced and the focus is shifted from one of looking for the right solution to one of identifying favorable Rationality Classes. Finally, we present a recursive framework that attempts to maximize prediction power under this new approach.

References and Notes

[1] i) Having as large a scope of validity as possible (the *Space of Cases of Interest* domain in Fig. 1); ii) Requiring low experimental effort (i.e., necessitating the least, or easiest to obtain, input information, while making comparable predictions); iii) Requiring low computational effort and; iv) Being as simple to interpret as possible are four other key properties of a good law. In this article we address only the issue of accuracy.

[2] Owen Gingerich. *The Eye of Heaven: Ptolemy, Copernicus, Kepler (Masters of Modern Physics)* (American Institute of Physics, New York, 1993).

[3] A. C. Ahn, M. Tewari, C.-S. Poon, R. S. Phillips. PLoS Medicine 3, e208 (2006).

[4] W. Rudin. *Real and Complex Analysis* (McGraw-Hill, 1987).

[5] Define the True Metric distance between two laws f and g as $\int_{All\,space} d(f(x), g(x)) d\mu$, where x is the Case variable, μ is the finite measure over the Space of Cases of Interest and d(f(x), g(x)) is the metric distance between f(x) an g(x) in the Space of Outcomes.

[6] Now the measure used in the Space of Cases of Interest assigns only nonzero point masses to the Cases with experimentally observed outcomes. As defined, this measure is appropriate only in a setting where the set of Case observations is finite.

[7] I. Mezić and T. Runolfsson. Automatica 44, 3003 (2008).

[8] G. Claeskens and N. L. Hjort. *Model Selection and Model Averaging* (Cambridge University Press. Cambridge, U.K., 2008).

[9] S. E. Baranzini. Autoimmunity 39, 651 (2006).

[10] S. H. Strogatz. Nature 410, 268 (2001).

[11] D. H. Zanette and S. C. Manrubia. Phys. Rev. Lett. 79, 523 (1997).

[12] The Observed Distances of laws are all computed using the same experimental data realization, so in general they are not independent. Two sets of laws could therefore have identical distributions of True Distance scores and yet have a different structure of correlations for the Observed Distances of their laws, hence the 'and Correlation Structure' clause.

[13] L. Preziosi. *Cancer Modeling and Simulation* (Chapman and Hall/CRC, London, 2003).

[14] S. A. Menchón and C. A. Condat. Phys. Rev. E 78, 022901 (2008).

[15] S. C. Ferreira, Jr., M. L. Martins and M. J. Vilela. Phys. Rev. E 65, 021907 (2002).

[16] B. Efron. *The Jackknife, the Bootstrap and Other Resampling Plans* (SIAM, Philadelphia, 1982).

[17] Biomarkers Definitions Working Group. Clin. Pharmacol. Ther. 69, 89 (2001).

[18] C. T. Keith, A. A. Borisy and B. R. Stockwell. Nature Reviews Drug Discovery 4, 71 (2005).

[19] J. F. Rual, K. Venkatesan, T Hao, et al. Nature 437 (7062): 1173-1178 (2005).

[20] J. P. Gonçalves, M. Grãos, and A. X. C. N. Valente. J. R. Soc. Interface 6 (39): 881-896 (2009).

[21] V. Trevino, F. Falciani and H. A. Barrera-Saldaña. Mol. Med. 13, 527 (2007).

[22] C. Boone, H. Bussey and B. J. Andrews. Nature Reviews Genetics 8, 437 (2007).

[23] A. X. C. N. Valente, J. A. B. Sousa, T. F. Outeiro, and L. Ferreira. arXiv 1003.1993v2 (2010).

[24] C. R. Scherzer, A. C. Eklund, L. J. Morse, et al. Proc. Nat. Acad. Sci. USA 104 (3): 955-960 (2007).

3

A Theory on Parkinson's Disease

Parkinson's Disease (PD) is considered essentially a pathology of nerve cells. Alterations beyond the nervous system are viewed as side-effects of the neurodegenerative disease. Differential gene expression in the blood of PD patients is deemed one such secondary effect and it is studied mostly from the perspective of enabling an early clinical diagnosis of the disease.

The article "A Stem-Cell Ageing Hypothesis on the Origin of Parkinson's Disease", proposes an alternative to the above conventional view. The Rationality Classes approach was applied to the analysis of genome-wide expression data from blood samples of Parkinson's patients and controls. The limitations of microarray based gene expression measurement and the relatively small dataset (105 net patients and controls) preclude definitive assertions. Nevertheless, the analysis clearly points to a broad number of functional processes known to be disrupted in Parkinson's Disease (PD) or other neurodegenerative processes also being, gene expression wise, altered in the blood cells of Parkinson's patients. This observation, coupled with a variety of results by other researchers in the field, led us to put forward two proposals regarding Parkinson's Disease, that we now summarize.

1. PD is best defined as a characteristic deviation from normality in the expression program of a cell. Blood cells can be in this PD expression state.

The observed altered gene expression in blood cells of Parkinson's patients is too comprehensive to be disregarded as a secondary event. Rather than focusing on albeit key singular manifestations of the disease, such as the formation of Lewy bodies and Lewy neurites, PD could be more deeply defined as a characteristic deviation from the normal expression program of a cell. We call it the PD expression state, or PD-state for short. The PD-state is a generic cell state. It is not specific only to nerve cells. In particular, blood cells can be in an integral PD-state. The critical role in neurons of some of the PD-state affected functional processes leads to catastrophic consequences specific to those cells, producing the observed neuronal-associated pathology.

2. The PD-state can originate as a case of hematopoietic stem cell program defect.

Early reports point to a time-scale on the order of years, for healthy neuronal grafts in PD patients to manifest pathological signs of PD, such as Lewy bodies and Lewy neurites. However, propagation from the nervous system to circulating blood cells would have to occur in accord with the much shorter blood turnover time-scale. This discrepancy, makes the direct nervous system to blood PD-state propagation route unlikely. Instead, we propose the blood PD-state originates as a case of hematopoietic stem cell program defect.

We further propose systemic propagation of the PD-state occurs over two distinct phases: A faster one through tissues under active renewal and a slower one through more stable tissues. Faster propagation to tissues under active stem cell based tissue regeneration would rely on the greater amenability of stem cells and their not yet fully differentiated progeny to reprogramming. Note this would signal an effective degree of communication between the hematopoietic stem cell niche and other adult stem cell niches in humans hitherto not considered. The propagation process would also benefit from the rapidity of the self-renewing process itself. Decrease in olfaction and gastrointestinal disturbances reported pre-motor signs of PD would be consistent with the high tissue turnover rates in the olfactory bulb and the gastrointestinal epithelia.

Our theory on Parkinson's Disease indicates checking whether the PD-state of a cell is locked in by epigenetic DNA modifications as a critical experiment to perform. Pondering hypotheses can elicit valuable directions for experimental efforts. At a time when it is becoming easier to acquire experimental data, this is as important as ever, lest there is a slide towards mere data collection with concomitant diminishing scientific returns.

A. X. C. N. Valente

4

Article

A Stem-Cell Ageing Hypothesis on the Origin of Parkinson's Disease

(arXiv: March 8th, 2010)

A Stem-cell Ageing Hypothesis on the Origin of Parkinson's Disease

André X. C. N. Valente[1,2], Jorge A. B. Sousa[2], Tiago Fleming Outeiro[3,4], Lino Ferreira[1,5]

[1]Center for Neuroscience and Cell Biology, University of Coimbra, Portugal

[2]Systems Biology Group, Biocant, Cantanhede, Portugal

[3]Cell and Molecular Neuroscience Unit, Instituto de Medicina Molecular, Lisboa, Portugal

[4]Instituto de Fisiologia, Faculdade de Medicina, Universidade de Lisboa, Portugal

[5]Biomaterials and Stem Cell Research Group, Biocant, Cantanhede, Portugal

A transcriptome-wide blood expression dataset of Parkinson's disease (PD) patients and controls was analyzed under the hypothesis-rich mathematical framework. The analysis pointed towards differential expression in blood cells in many of the processes known or predicted to be disrupted in PD. We suggest that circulating blood cells in PD patients can be in a full-blown PD-expression state. We put forward the hypothesis that sporadic PD can originate as a case of hematopoietic stem cell/differentiation process expression program defect and suggest this research direction deserves further investigation.

Introduction

Parkinson's disease (PD) is the second most common neurodegenerative disorder, after Alzheimer's disease [1]. With the increasing life-expectancy, the number of worldwide affected individuals is expected to double from ~5 million in 2005 to ~8 million by 2030 [2]. The majority of PD cases are sporadic, with only 5-10% of cases presumed to have a well-defined singular genetic cause. In spite of significant insights and hypotheses proposed in recent years, the origin (etiology) of sporadic PD remains undetermined [1].

A landmark study has reported genome-wide expression data on RNA extracted from whole blood of 50 predominantly early-stage PD patients (mean Hoehn and Yahr stage 2.3, range 1-4) and 55 age-matched controls [3]. The study data analysis focused on producing a candidate blood-based early diagnostic composite biomarker for PD,

45

but not on any potential implications of the data towards a renewed biological understanding of PD and its etiology. In fact, many of the genes it identified as blood PD biomarkers have no known role in PD pathogenesis [3].

Recently, we developed a novel mathematical framework, named hypothesis-rich framework, for analyzing problems that are complex in the sense of involving a large number of a priori potentially relevant hypothesis, facts and data [4]. The framework is ideal for addressing systems biology problems where there is potential to jointly leverage a diversity of existing biological knowledge and quantitative data on a large number of variables. Having significant prior knowledge, which has accumulated through the vast work on Parkinson's disease (PD) until the present day, and having a large number of biological variables in the 22 283 gene expression probe readings, the hypothesis-rich framework is well suited for profiling and interpreting the PD blood gene expression data.

Herein, we analyzed the PD blood gene expression dataset previously reported, now under the hypothesis-rich mathematical framework. We constructed a composite biomarker, based on the gene expression data, for distinguishing PD individuals from non-PD individuals. The composite biomarker showed an across-the-board presence of genes known to play a role in PD and neurodegenerative processes. We propose this raises the prospect of an integral presence of PD in the blood. We hypothesize that a hematopoietic stem cell defect can be at the origin of sporadic PD.

Results

In settings where there are large number of candidate biomarker laws, noisy data and a limited number of observed cases, a fundamental obstacle arises in that many candidate laws will diagnose correctly the observed cases just by chance. This is the so called over-fitting, or under-determination issue. Our PD classifier problem falls in this regime: i) we are searching for a law potentially based on 22 283 probe readings; ii) we have data on just 105 patients and iii) noise can be a significant issue in microarray experiments [5]. The hypothesis-rich approach to cope with this fundamental obstacle consists in leveraging biological knowledge and insights to group candidate laws into Rationality Classes (RCs). Separating candidate laws into RCs enhances the statistical ability to circumvent over-fitting, increasing the chance of finding good laws, by way of finding favorable RCs.

Technically, this stems from different RCs having different True Distance Distribution and Correlation Structures, due to their distinct etiology and rationale [4]. We refer the reader to another article [4] for the underlying theory and report here in *Methods* only how the technique was implemented in this PD biomarker problem.

The blood gene expression-based final composite PD biomarker law is shown in Table 1. The biomarker demonstrated a positive, predictive ability with a Receiver Operating Characteristic (ROC) area under the curve (AUC) of 0.59 ± 0.07 (see *Methods*). Interestingly, most of the 14 laws that compose the PD biomarker turned out to be centered around genes previously associated with PD and/or other neurodegenerative pathologies:

1. CLTB: This gene, encoding clathrin, is associated with vesicle-mediated transport, especially in endocytosis. CLTB has been implicated in dopamine transporter endocytosis [6] and shown in a recent PD mouse model to be down-regulated in the striatum region of the brain [7]. Both our biomarker and that of Scherzer et al. [3] associate high expression of CLTB in the blood with the presence of PD.

2. FPR3: This gene belongs to the formyl peptide receptor family (FPR1, FPR2, and FPR3) of G-protein coupled receptors. FPRs have been implicated in the pathogenesis of Alzheimer's disease (AD) and prion diseases [8]. In addition, FPRs can activate microglia [9] which has been observed in brains from patients with PD [10]. Both our analysis and that of Scherzer et al. [3] found FPR3 to be over-expressed in the blood of PD patients.

3. ITGA2*ITGB1*CD47 [Integrin Complex [11]]: Integrins have been shown to mediate the interaction between microglia, the resident macrophages of the Central Nervous System, and the fibrillar beta-amyloid plaques found in the brains of Alzheimer's disease patients [12]. In our analysis, high values of the gene expression product of the integrin complex biomarked the Alzheimer's and other neurodegenerative diseases (A) group (vs. the Parkinson's disease (P) group and the healthy (H) group).

4. UBOX5*PRPF19: PRPF19 and UBOX5 have been described to have ubiquitin ligase activity [13]. A high-throughput data analysis by Hourani et al. [14] included PRPF19 in a set of 35 genes differentially expressed in the left brain hemisphere of mice with induced PD compared to normal controls. We are not aware of studies specifically associating UBOX5 and neurodegenerative processes, beyond its

a)

	Functional group	One-sided biomarker type	Expression sign	Genes and mathematical formula	Affymetrix HG-U133A probes
1	Vesicle-mediated transport	P vs. (A & H)	High in P	CLTB	211043_s_at
2	G-protein coupled receptors	P vs. (A & H)	High in P	FPR3	214560_at
3	Human literature protein complexes	A vs (P & H)	High in A	CD47 * ITGA2 * ITGB1	(211075_s_at * 213055_at * 213856_at * 213857_s_at) * (205032_at) * (211945_s_at * 215878_at * 215879_at * 216178_x_at * 216190_x_at)
4	Ubiquitination	A vs (P & H)	Low in A	UBOX5 * PRPF19	215544_s_at * 203103_s_at
5	Vesicle-mediated transport	P vs. (A & H)	High in P	CLTB * KIF5B	211043_s_at * 201991_s_at
6	Inflammatory	H vs. (A & P)	Low in H	PTGS2 * CYBB	204748_at * 217431_x_at
7	Oxidative stress	A vs (P & H)	Low in A	GPX4	201106_at
8	Dopamine	A vs (P & H)	High in A	PTGS2	204748_at
9	Genetically linked to Parkinson	H vs. (A & P)	Low in H	UCHL1	201387_s_at
10	Folding	H vs. (A & P)	High in H	ST13	208666_s_at
11	Nitric Oxide	A vs (P & H)	High in A	ACTB	AFFX-HSAC07/X00351_M_at
12	Literature physically interacting protein pairs	H vs. (A & P)	High in H	SMAD3 * STRAP	205398_s_at * 200870_at
13	Ubiquitination	P vs. (A & H)	High in P	UBOX5 * CUL4B	215544_s_at * 202213_s_at
14	Ubiquitination	H vs. (A & P)	High in H	SMAD3	205398_s_at

b)

Parkinson's disease composite final law
+ CLTB / 55.7
+ FPR3 / 28
+ min (- CD47*ITGA2*ITGB1, - 1.806*10^{18}) / (3.131*10^{18})
+ min (UBOX5*PRPF19, 8023.4) / 5916.2
+ CLTB*KIF5B / 16654
+ min (PTGS2*CYBB, 1109.5) / 1327.6
+ min (GPX4, 544.1) / 124.9
+ min (- PTGS2, - 68.1) / 36.2
+ min (UCHL1, 81.2) / 42.5
+ min (- ST13, - 90.5) / 25.4
+ min (- ACTB, - 5761.7) / 1420.9
+ min (- SMAD3*STRAP, - 16611.8) / 8477.5
+ UBOX5*CUL4B / 4221.5
+ min (- SMAD3, - 50.5) / 24.9

Table 1. a) The 14 laws that make up the final composite Parkinson's disease biomarker and their respective Rationality Class of origin. **b)** The composite gene-expression biomarker for Parkinson's disease. Gene symbols stand for their expression level. For normalization, every term is divided by its standard deviation.

generic ubiquitin ligase activity [15]. Our analysis shows this gene expression product as being under-expressed in the A group (vs. the P and H groups).

5. CLTB*KIF5B: KIF5B, a member of the kinesin family, is associated with the transport of peptide-containing vesicles to neuron terminals [16]. In a genome-wide expression profiling of substantia nigra dopamine neurons in PD patients vs. controls, KIF5B was identified as one of the PD differentially under-expressed genes [16]. CLTB has been discussed above. Our biomarker associates high values of this gene expression product with the presence of PD.

6. PTGS2*CYBB: PTGS2 encodes the inducible form of cyclooxygenase (COX-2). Teismann et al. demonstrated that PTGS2 inhibition averts the formation of the oxidant species dopamine-quinone, which has been implicated in the pathogenesis of PD [17]. A genetic association study showed that polymorphisms of the PTGS2 gene predispose to AD [18]. In mice over-expressing the Swedish mutation of the amyloid precursor protein as a model of AD, it was found that lack of CYBB averted the development of oxidative stress, cerebro-vascular dysfunction and behavioral deficits [19]. This gene expression product showed lower values in the H group vis-à-vis in the P and A groups.

7. GPX4: This gene encodes glutathione peroxidase 4 and is involved in inflammation and in the oxidative stress response [20]. GPX4 is a target gene of DJ-1 [21], a protein associated with familial cases of PD [22]. Blackinton et al., although not detecting changes in mRNA expression levels of GPX4, observed an increase in its protein expression levels in the cortex of PD patients vs. controls [21]. GPX4 is also a key antioxidant against lipid peroxidation, which was shown to be an early event in AD [23]. Our analysis showed GPX4 being expressed at lower levels in the A group.

8. PTGS2: Already discussed above. Our analysis showed PTGS2 being expressed at higher levels in the A group.

9. UCHL1: Mutations in UCHL1 are associated with familial cases of PD [24]. UCHL1 hydrolyzes a peptide bond at the C-terminal glycine of ubiquitin. It was found to be decreased in the H group in our analysis.

10. ST13: The protein encoded by this gene is a cofactor of chaperone heat-shock protein HSP70 [25]. Depletion of HSP70 has been associated with the formation of Lewy bodies in PD patients [26]. In vitro and in vivo experiments, the latter using Caenorhabditis

elegans, indicate that a lower expression level of ST13 could facilitate depletion of HSP70 [27]. Both our biomarker and that of Scherzer et al. [3] associate high expression levels of ST13 in the blood with the H group.

11. ACTB: This gene, encodes beta-actin. Actins are highly conserved proteins involved in cell structure and motility [28]. Beta-actin, in particular, was found to regulate platelet nitric-oxide synthase 3 activity [29]. There is evidence for the presence of Beta-actin mRNAs within developing dendritic and axonal growth cones [30]. A high-throughput study compared gene expression in cerebral cortices of AD patients with that in cerebral cortices of non-demented controls containing abundant amyloid-plaques [31]. It found ACTB to be the second most differentially over-expressed gene in the AD patients. However, at the time, follow-up RT-PCR analysis did not corroborate this differential expression. Our biomarker associates high expression levels of ACTB with the A group.

12. SMAD3*STRAP: SMAD3 and STRAP are involved in transforming growth factor β signaling. (TGF-β) [32]. TGF-β levels have been reported as elevated in the striatum and in ventricular cerebrospinal fluid in PD [33,34]. In the blood, this gene expression product showed higher values in the H group.

13. UBOX5*CUL4B: UBOX5 and CUL4B have been described to have ubiquitin ligase activity [13]. We are not aware of studies specifically associating UBOX5 or CUL4B and PD, beyond the general fact that the ubiquitin proteasome system has been widely implicated in protein accumulation in neurodegeneration [15]. In the blood, this gene expression product showed higher values in the P group.

14. SMAD3: Already discussed above. Our analysis showed SMAD3 being expressed at higher levels in the H group.

Discussion

From the standpoint of serving as an early PD diagnostic tool, our composite biomarker demonstrated a lower predictive ability than that of the biomarker proposed and evaluated by Scherzer et al. (AUC of 0.69) [3]. Although at present noise remains a significant factor in genome-wide expression measurements [5], there is intensive ongoing research to address this limitation, centered both on microarray technology and on emerging techniques such as deep sequencing [35]. Whether reductions in such technical noise will reveal any of these PD blood biomarkers to be powerful enough for the practical diagnosis of PD, or whether their limited predictive power is intrinsically biological, is an open question to be answered in the coming years. However, most interestingly, our blood gene expression biomarker already shows a significant, across-the-board presence of genes and processes known to play a role in PD and in neurodegenerative processes. Differential expression in blood cells in many of the processes known or predicted to characterize PD in neurological tissues, opens the possibility that, more than just exhibiting side-effects of the presence of PD in neuronal tissues, circulating blood cells in PD patients may be, at least expression-wise, in a full state of PD, much in the same way as affected neurons are. In fact, several studies, when taken as a whole, give further credibility to this possibility. Cytoplasmic hybrid ("cybrid") models of PD, in which donor mtDNAs from PD patients' platelets are introduced into and expressed in neural tumor cells with identical nuclear genetic and environmental backgrounds, demonstrate many abnormalities in which increased oxidative stress drives downstream antioxidant response and cell death activating signaling pathways [36]. PD cybrids spontaneously form Lewy bodies and Lewy neurites, linking platelet mtDNA expression to neuropathology [37]. Furthermore, they show both reduced mitochondrial respiration and impaired organelle transport in processes [36,38]. Thus, PD cybrids demonstrate that mitochondria from at least one blood cell type, contain the gene expression programs which are necessary for the expression of several central features of the disease. In addition, peripheral blood lymphocytes (PBL) from PD patients display altered densities of D1 and D2 dopamine receptors [39], further supporting our hypothesis. We believe the possibility of an integral presence, at least expression-wise, of PD in circulating blood cells should be considered. It could have deep implications on the open question of the etiology of sporadic PD (the word sporadic will be subsumed henceforth) and we discuss these next.

One commonly held generic view is that an environmental agent (or agents), possibly exacerbated by genetic and/or age-related vulnerabilities, is at the root of PD [40]. Proposed candidate agents include environmental poisons, such as pesticides and metals [40], and yet unidentified neurotropic pathogens, such as viruses [41] or a prion-like protein [42]. Usually suggested points of induction, or routes of entry for such a pathogen, are the peripheral olfactory system [42] and the gastrointestinal tract [43], due to the known association of these systems with early clinical symptoms and also Lewy body pathology in the latter case. These two candidate routes have recently been combined into a dual-hit theory [41]. The theory proposes the pathogen, possibly a virus, would initially enter the body via the nasal route, and then via being swallowed in saliva and mucus, cross the stomach wall. Oxidative stress effects spur another major research direction on the origin of PD [44]. Finally, regardless of the ultimate trigger, the criticality of the neuro-inflammatory response in the PD onset process is also a matter of active debate [45].

Taking our observation as a starting point, we put forward an alternative hypothesis on the nature and etiology of PD (Fig.1). We first suggest that, at its root, PD is best defined as a characteristic deviation from normality in the expression program of cells. We call it the PD-expression state, or PD-state for short. Thus, we propose that circulating blood cells in PD patients are in a full PD-state. The PD-state would therefore be a generic cell state, not specific to neuronal cells. However, as also argued by others [46], due to the particular critical role that some expression programs play in neuronal tissues, the PD-state is catastrophic in them, leading to the observed neuronal-associated pathology. A crucial question then becomes whether the PD-state is propagated from neuronal cells to blood cells, or vice-versa. That is, where does it originate? Recent studies show PD pathological signs, such as Lewy bodies and Lewy neurites, being propagated to healthy neuronal grafts in PD patients only over a time-scale on the order of a decade [47,48]. Considering, by comparison, the much shorter life-time of most blood cells (the lifespan of red blood cells and platelets is 127 [49] and 4.4 [50] days, respectively), we argue that it is comparatively more realistic that the PD-state originated in the blood cells.

PD is markedly age associated, with only 4% of PD cases diagnosed in the United States occurring before age 50 [2]. There is evidence that hematopoietic stem cells (HSCs) age, showing an altered cell surface phenotype and changes in metabolic activity and gene expression [51,52]. Recent studies demonstrated that this ageing process is a consequence of accumulation of DNA damage [53]. These lesions can be propagated to daughter stem cells and to downstream lineages through the processes of self-renewal and differentiation. We propose that the PD-state acquired by blood cells could be a case of hematopoietic stem cell ageing. Under this premise, circulating endothelial progenitor cells, which undergo endothelial cell differentiation under appropriate inductive signals and form neovessels [54], become a candidate vehicle for propagation of the PD-state to other cells in the human body.

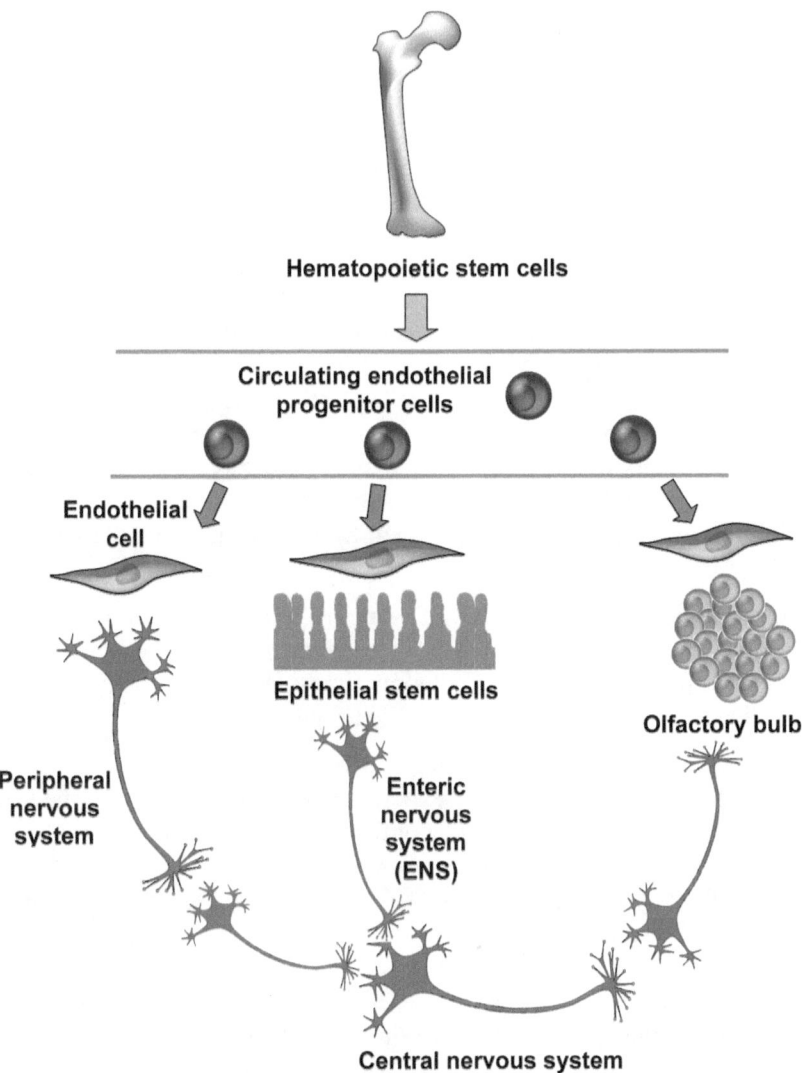

Figure 1. Bone marrow-derived stem cells as the origin of PD. According to this hypothesis, the PD-state originally appears in bone marrow-derived stem cells due to ageing. Bone marrow-resident hematopoietic stem cells give rise to blood cells in a PD-state, including circulating endothelial progenitor cells (CEPCs). CEPCs differentiate into endothelial cells which are incorporated into blood vessels, hence propagating the disease at different sites of the human body. The first symptoms of the disease (as early as 10 years before motor symptoms) would occur at places where the cell turnover is high, for instance at gastrointestinal tissue and at the olfactory bulb.

53

An early sign of PD is impaired sense of smell [42]. The olfactory bulb was recently reported as being a site of continuous stem cell based tissue regeneration [55]. Gastrointestinal dysfunction is another early sign of PD reported as much as 10 years before motor symptoms appear [56]. Gastrointestinal function is very sensitive to the proper function of intestinal epithelial cells [57], which are replenished by local adult stem cells with tissue turnover in under 7 days [58]. Assuming the ability of HSCs to propagate the PD-state to adult stem cells via circulating progenitor cells, would explain the early PD symptoms in sites of very active stem cell based tissue regeneration. The initial propagation of the PD-state to these tissues could be due to rapid self-renewing and stem cell plasticity facilitating stem cell reprogramming by the endothelial cells derived from the circulating progenitor cells [59,60]. Of note, α-synuclein has been shown to be expressed in endothelial cells [61]. Alternatively, the niche stem cells might be directly under replenishment by transformed endothelial cells [62].

Our proposition is that the PD-state is initially disseminated in a shorter time-frame through the differentiation process of active stem cell niches. Then, over a distinct slow time-scale on the order of years, the PD-state propagates through stable tissues.

We believe that the hypothesis presented here deserves further investigation and that some experiments should be performed to validate this line of research. To confirm the involvement of circulating progenitor cells in the propagation of PD-state, CD34+ cells (or subpopulations of CD34+ cells) [63] collected from the peripheral blood of PD patients could be transplanted in the bone marrow of nude mice in which all the bone marrow cells are ablated by irradiation prior to the transplant. This experiment might give important insights on whether circulating progenitor cells are involved in the etiology of PD. To evaluate whether PD originates at the bone marrow and propagates to the nervous system, or rather initiates at the gastrointestinal tissue, propagates to the nervous system and only via this latter one, reaches the bone marrow, we suggest a long-term, large-scale study where the blood of individuals with ages above 50 years but without PD symptoms would be collected and analyzed over 10 years. If the bone marrow is implicated in the origin of PD, circulating progenitor cells will exhibit a PD-state profile before motor symptoms appear. In contrast, if motor symptoms appear before the PD-state profile is observed in circulating progenitor cells, then the bone marrow is not the origin but rather yet another late stop in the progression of the disease. We chose to highlight one particular hypothetical path from stem cell ageing to PD, but variant paths cannot be excluded at present. For instance, PD could also originate in the transformation (ageing) of intestinal epithelial stem cells (Lgr5$^+$ cells). Of note, these stem cells have been recently identified as the origin of intestinal cancer [64]. In this regard, the additional collection and characterization of intestinal biopsies from the individuals in the aforementioned proposed study would be pertinent to discriminate between these alternate possibilities.

Methods

In *Basic algorithm* we describe the overall procedure for building our composite biomarker, based on the testing of RCs. In *Rationality Classes for Parkinson's disease* we describe the RCs that we constructed for this PD problem.

Basic algorithm

Henceforth, P shall refer to "Parkinson's", H shall refer to "healthy" and A shall refer to "Alzheimer's or other neurological disease". Define a law as a rule that assigns a value to each individual. For example, the expression value of gene GeneX on the individual is a law. The product of the expression values of GeneX, GeneY and GeneZ is another example of a law. The final objective will be to select a law whose values constitute a good diagnostic score for the presence of P. We shall use the convention of equating high values with higher chance of P.

We define the following Law Scoring Function (LSF) to rate how good a law is, based on a set of individuals:

$$\text{LSF score} = \left| \frac{1}{\dfrac{\sigma_P}{<P> - <AH>} + \dfrac{\sigma_{AH}}{<P> - <AH>}} \right|,$$

where,

$<P>$ = the average [law value] over the P individuals in the set,

$<AH>$ = the average over the H and A individuals in the set,

σ_P = the standard deviation over the P individuals in the set,

σ_{AH} = the standard deviation over the H and A individuals in the set.

The LSF score measures, for the set of individuals in question, how well separated is the mean of the P individuals from the mean of the AH individuals, both in terms of the standard deviation of the P individuals and in terms of the standard deviation of the AH individuals.

Candidate laws are grouped in RCs, as described in the next section. Based on these RC sets of laws, we now describe in steps how to select a final law:

1. The available set of individuals is randomly divided 40/60 into two sets, set S1 and set S2.

2. The laws in each RC are scored and ranked using the LSF and set S1.

3. A new Super Set of laws is assembled containing the top N ranked laws from each RC [note: due to the small size of the overall dataset in question, it was not possible to further generate a meaningful internal procedure for optimizing N. N=10 was used as a reasonable a priori choice].

4. The laws in the new Super Set are scored and ranked using the LSF and the unused set S2 of patients.

5. A composite law is built. It is built by additive combination of laws to be selected from the top N ranked laws in the Super Set. This selection to belong to the composite law is done as follows: Starting with the top ranked law in the Super Set, and moving down the rank, we test whether adding the next law in question to the composite law increases the LSF score of the current composite law. If it does, then the law is selected and the composite law is updated. In the above process, a law is first normalized via division by its standard deviation in the S2 set. Now, each RC has an RC-counter that counts how many times that RC has contributed with a law to the composite law. Each time a RC contributes with a law, its RC-counter increases by one. The number of members in the final composite law is also recorded. Note that when adding laws as described, care must be taken to add them with the appropriate sign, so that high values always respect the convention of being associated with P.

6. Steps 1 through 5 are repeated, for a new random 40/60 partition into S1 and S2 sets. The RC-counters increase cumulatively throughout these repetitions. The 1 through 5 cycle is repeated until the RCs can be ranked with confidence by their RC-counters (i.e., the ranking approaches the unique ranking in the limit of infinite iterations). The average, over these repeated cycles, of the number of members in the final law is also recorded.

The purpose of the above steps was exclusively to obtain i) the RC-counters ranking of the RCs and ii) the average number of members in the final law. We now turn to actually constructing the final composite law:

7. The RC-counters are all scaled by a common factor. This factor is the largest small enough factor such that after the scaling followed by a subsequent rounding to the nearest integer, the sum of the RC-counters is no greater than the average number of members in the final law, as computed above. The number of laws that RCs contribute to the final

composite law are then given by these new, scaled and rounded to the nearest integer, RC-counter values.

8. Using this time the full available set of individuals, the laws in each contributing RC are LSF scored and ranked accordingly. Each RC then contributes to the final composite law the top ranked number of laws determined in step 7 above. Similarly to what occurs in step 5, laws are first normalized via division by their standard deviation, this time in the full set of available individuals. The final composite law is thus built.

To the above procedure there is one significant modification to accommodate what we call one-side biomarkers. This is described next.

One-sided biomarkers

Although the final aim is to build a P vs. non-P classifier, a law that distinguishes well a subset of P individuals from the (rest of P individuals U non-P individuals) would be valuable, as at least it would allow a subset of the P individuals to be identified. Likewise, a law that distinguished well a subset of non-P individuals from the (P individuals U rest of the non-P individuals) set, would be useful too, as it would allow a subset of the non-P individuals to be identified. We call such laws one-side biomarkers, as only on one side of their scale of values do they provide a conclusive diagnostic. We suggest that for complex, heterogeneous pathologies, one-sided biomarkers may turn out to be crucial, beneficially bio-marking and characterizing those pathologies via a diversity of sub-cases. In studies involving a non-extensive number of individuals searching for one-sided biomarkers is difficult, due to insufficient statistical power. However in this P problem we have the advantage of the non-P group being naturally decomposable into two already identified, distinct groups: the H group and the A group.

To the P vs. (AH) type of biomarker, we add two types of one-sided biomarkers: The H vs. (AP) one-sided biomarker type and the A vs. (HP) one-sided biomarker type. Our bookkeeping convention will be to associate a biomarker type to a RC. The algorithm shown in steps above is then modified as follows to accommodate this change:

i) In RCs of type H vs. (AP) the LSF to be used is

$$\text{LSF score} = \left| \frac{1}{\dfrac{\sigma_{AP}}{<AP> - <H>} + \dfrac{\sigma_H}{<AP> - <H>}} \right|,$$

where,

    $\langle AP \rangle$ = the average over the A and P individuals in the set,

    $\langle H \rangle$ = the average over the H individuals in the set,

    σ_{AP} = the standard deviation over the A and P individuals in the set,

    σ_{H} = the standard deviation over the H individuals in the set.

Similarly, in RCs of type A vs. (HP) the LSF to be used is

$$\text{LSF score} = \left| \frac{1}{\dfrac{\sigma_{HP}}{\langle HP \rangle - \langle A \rangle} + \dfrac{\sigma_{A}}{\langle HP \rangle - \langle A \rangle}} \right|,$$

where,

    $\langle HP \rangle$ = the average over the H and P individuals in the set,

    $\langle A \rangle$ = the average over the A individuals in the set,

    σ_{HP} = the standard deviation over the H and P individuals in the set,

    σ_{A} = the standard deviation over the A individuals in the set.

ii) In step 3 now three Super Sets are built, each drawing from RCs of one of the above three types.

iii) In step 5, the composite final law is built by testing in turn a law from each of the three internally rank ordered Super Sets. Through this process, the top N laws from each Super Set will be tested. Now, consider the A vs. (HP) one-sided biomarker. For purposes of our ultimate objective of distinguishing P vs. non-P individuals, this one-sided biomarker is of diagnostic value only on one side of its scale of values, namely on the side identified with A. Suppose the low values are the ones associated with A, while the high values are inconclusive, as they are associated with both H and P. Then, when adding (or testing the addition of) this law to the composite law, rather than adding its value V, we add min (V, midpoint), where midpoint = $(\langle A \rangle + \langle HP \rangle)/2$ is the midpoint between the averages of the A and (HP) groups. This way, as it should, only the left side of the scale is of significance in the P vs. non-P composite law. If high values were the ones associated with A, then we should add min(-V, -midpoint), in keeping with our convention of having high values associated with P. Analogous arguments apply to H vs. (AP) biomarkers and to building the final composite law in step 8.

Rationality Classes for Parkinson's disease

We constructed RCs by combining three different factors (Supp. Mat. Figure 1):

1. Group type: First, gene ontology based [13] functional sets of genes were chosen according to existing knowledge or existing predictions that they may play a relevant role in PD. A set of 9 genes that have been genetically linked to PD was also considered. To these, a basic set containing all genes and a set containing only the roughly 1000 most highly expressed genes were added. Finally, protein physical interactions were also considered [65], via two sets of human protein-protein pair-wise interactions (one based on literature curated data [66], the other based on yeast-to-hybrid high-throughput data [66,67]), and a set of literature curated human protein complexes [11].

2. Biomarker type: This could be the P vs. (AH) standard biomarker or the A vs. (HP) or H vs. (AP) one-sided biomarkers.

3. Mathematical formula: Laws based on single entries, the product of a pair of entries and the log of the ratio of two entries were considered. Further, the single and product cases were both further subdivided into two cases, according to whether the high or the lower value of the law is the one associated with P.

By combining the above three factors, 291 RCs were constructed. For example, the set of laws that try to distinguish A from (PH), associate single gene high values with (PH) and where the single gene belongs to the vesicle-mediated transport functional set constitutes one RC.

Remarks:

1. For protein binary interactions only the product and the log ratio of the interacting pair are considered.

2. The entry associated with a protein complex is the product of the expression of its member genes (note this allows products and log ratios amongst complexes to be also considered).

3. For large functional sets, we only evaluated products and log ratios amongst the 200 entries in the set that had the highest LSF scores as single entries. This was for computational time reasons.

4. The highly expressed gene set is not a priori determined and held fixed. This is so in order to maintain proper statistical independence within the algorithm. Hence why we state above that its number of genes is 1000 only approximately.

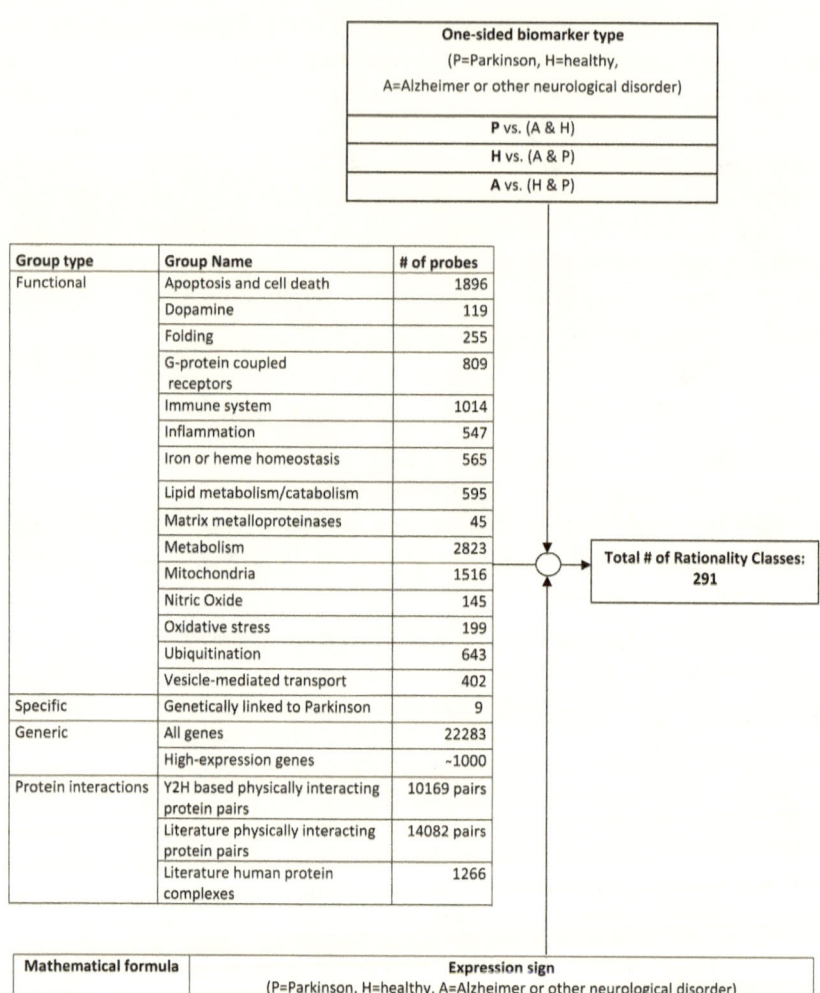

Supp. Mat. Figure 1. The construction of Rationality Classes for Parkinson's disease genome-wide expression analysis.

Notes

Computation of the Final Composite Law and the associated area
under the curve (AUC) with confidence interval

When computing the final composite law, 5 000 iterations of the step
1 through step 5 cycle in the algorithm description above were
performed. The top 50 ranked by RC-counter RCs are shown in Supp.
Mat. Table 1. The number of patients in the sets S1 and S2 also alluded
to in the algorithm description above was 42 and 63, respectively.

To determine the associated area under the curve (AUC), an extra
external cycle, leaving 5 random patients out at time, was created. The
number of patients in the sets S1 and S2 became 40 and 60,
respectively. The outer testing cycle was repeated 500 times and the
iterations of the inner step 1 through step 5 cycle were reduced to 50,
for computational time reasons.

When determining the standard deviation confidence interval for the
AUC estimate, two approximations were made. First, liberally, we
assume we have approximately tested 105 independent patients.
Second, conservatively, we consider only the minimum number of
guaranteed independent (PD patient, non-PD patient) pairs amongst
these 105 patients [68,69]. This minimum number is $\min(50,55)=50$,
since there are 55 PD patients and 50 non-PD patients. The standard
deviation in the AUC estimate thus becomes $\text{sqrt}[AUC*(1-AUC)/50] \leq$
$\text{sqrt}[0.25/50] \approx 0.07$.

Datasets

The blood genome-wide expression dataset on Parkinson's patients
and controls was obtained from Scherzer et al. [3]. The data was used
as provided in the NCBI Geo Gene Expression Omnibus [70], without
any further normalization. The literature curated human protein-protein
pair-wise interactions dataset was obtained from Rual et al. [66]. The
Y2H based high-throughput human protein-protein pair-wise
interactions dataset consists of the union of the Rual et al. [66] and the
Stelzl et al. [67] datasets. The literature curated human protein
complexes dataset was obtained from Ruepp et al. [11].

	Class Name	Class Type	Expression Sign	Operation Type	RC-counter (after 5000 iterations)
1	Vesicle-mediated transport	P vs. (A & H)	High in P	Simple	2880
2	G-protein coupled receptors	P vs. (A & H)	High in P	Simple	1989
3	Literature human protein complexes	A vs (P & H)	High in A	Simple	1642
4	Ubiquitination	A vs (P & H)	Low in A	Product	1624
5	Vesicle-mediated transport	P vs. (A & H)	High in P	Product	1390
6	Inflammation	H vs. (A & P)	Low in H	Product	1345
7	Oxidative stress	A vs (P & H)	Low in A	Simple	1296
8	Dopamine	A vs (P & H)	High in A	Simple	1232
9	Genetically linked to Parkinson	H vs. (A & P)	Low in H	Simple	1182
10	Folding	H vs. (A & P)	High in H	Simple	1137
11	Nitric Oxide	A vs (P & H)	High in A	Simple	1126
12	Literature physically interacting protein pairs	H vs. (A & P)	High in H	Product	1101
13	Ubiquitination	P vs. (A & H)	High in P	Product	976
14	Ubiquitination	H vs. (A & P)	High in H	Simple	975
15	Nitric Oxide	A vs (P & H)	Low in A	Simple	972
16	Immune system	H vs. (A & P)	High in H	Simple	957
17	Literature human protein complexes	H vs. (A & P)	High in H	Simple	932
18	Oxidative stress	A vs (P & H)	High in A	Simple	905
19	Dopamine	H vs. (A & P)	Low in H	Simple	885
20	Iron or heme homeostasis	H vs. (A & P)	Low in H	Product	872
21	Ubiquitination	H vs. (A & P)	High in H	Product	852
22	Dopamine	A vs (P & H)	High in A	Product	846
23	High-expression genes	A vs (P & H)	Low in A	Simple	758
24	High-expression genes	P vs. (A & H)	High in P	Simple	724
25	Literature physically interacting protein pairs	H vs. (A & P)	High absolute	Ratio Log	721
26	Folding	H vs. (A & P)	High absolute	Ratio Log	665
27	Genetically linked to Parkinson	P vs. (A & H)	High in P	Simple	659
28	Iron or heme homeostasis	A vs (P & H)	High in A	Simple	651
29	High-expression genes	A vs (P & H)	Low in A	Product	599
30	Iron or heme homeostasis	A vs (P & H)	High absolute	Ratio Log	596
31	Inflammation	A vs (P & H)	Low in A	Simple	566
32	All genes	P vs. (A & H)	High in P	Simple	560
33	Mitochondria	H vs. (A & P)	High in H	Product	555
34	Oxidative stress	H vs. (A & P)	Low in H	Simple	531
35	Literature human protein complexes	P vs. (A & H)	Low in P	Simple	507
36	Iron or heme homeostasis	A vs (P & H)	High in A	Product	500
37	G-protein coupled receptors	P vs. (A & H)	High in P	Product	499
38	Ubiquitination	A vs (P & H)	Low in A	Simple	483
39	Apoptosis and cell death	H vs. (A & P)	High absolute	Ratio Log	436
40	Mitochondria	P vs. (A & H)	High absolute	Ratio Log	431
41	High-expression genes	P vs. (A & H)	High in P	Product	428

42	Literature human protein complexes	H vs. (A & P)	Low in H	Simple	407
43	Matrix metalloproteinases	P vs. (A & H)	Low in P	Product	397
44	Nitric Oxide	A vs (P & H)	Low in A	Product	368
45	Inflammation	A vs (P & H)	Low in A	Product	364
46	Mitochondria	P vs. (A & H)	High in P	Simple	361
47	Folding	P vs. (A & H)	High absolute	Ratio Log	359
48	Apoptosis and cell death	A vs (P & H)	Low in A	Product	354
49	Mitochondria	H vs. (A & P)	High in H	Simple	347
50	Ubiquitination	P vs. (A & H)	High in P	Simple	342

Supp. Mat. Table 1. The top 50 Rationality Classes, rank ordered by RC-counter.

Go annotations

The functional groups used when constructing Rationality Classes were based on the Gene Ontology classification scheme [13], as given in the Affymetrix Human Genome U133 Array data sheet of November 2009.

Acknowledgements

TFO is supported by an EMBO installation grant, a Marie Curie International Reintegration Grant and FTC Grant PIC/IC/82760/2007.

References

[1]   Davie CA. A review of Parkinson's disease. Br Med Bull. 2008: p. 109-27.

[2]   Dorsey ER, Constantinescu R, Thompson BA, Biglan KM, Holloway RG, Kieburtz K, et al. Projected number of people with Parkinson disease in the most populous nations, 2005 through 2030. Neurology. 2007: p. 384-6.

[3]   Scherzer CR, Eklund AC, Morse LJ, Liao Z, Locascio JJ, Fefer D, et al. Molecular markers of early Parkinson's disease based on gene expression in blood. Proc Natl Acad Sci U S A. 2007: p. 955-60.

[4]   Valente AXCN. Prediction in the hypothesis-rich regime. arXiv: 1003.3551v1. 2010.

[5]   Tu Y, Stolovitzky G, Klein U. Quantitative noise analysis for gene expression microarray experiments. Proc Natl Acad Sci U S A. 2002: p. 14031–6.

[6]   Sorkina T, Hoover BR, Zahniser NR, Sorkin A. Constitutive and protein kinase C-induced internalization of the dopamine transporter is mediated by a clathrin-dependent mechanism. Traffic. 2005: p. 157-70.

[7]   Zhang X, Zhou JY, Chin MH, Schepmoes AA, Petyuk VA, Weitz KK, et al. Region-specific protein abundance changes in the brain of MPTP-induced Parkinson's disease mouse model. J Proteome Res. 2010: p. 1496-509.

[8]   Lorton D, Schaller J, A. L, De Nardin E. Chemotactic-like receptors and ABeta peptide induced responses in Alzheimer's Disease. Neurobiol Aging. 2000: p. 463–73.

[9]   Gao X, Hu X, Qian L, Yang S, Zhang W, Zhang D, et al. Formyl-methionyl-leucyl-phenylalanine–Induced Dopaminergic Neurotoxicity via Microglial Activation: A Mediator between Peripheral Infection and Neurodegeneration? Environ Health Perspect. 2008: p. 593–8.

[10] McGeer PL, Itagaki S, Boyes BE, McGeer EG. Reactive microglia are positive for HLA-DR in the substantia nigra of Parkinson's and Alzheimer's disease brains. Neurology. 1988: p. 1285-91.

[11] Ruepp A, Waegele B, Lechner M, Brauner B, Dunger-Kaltenbach I, Fobo G, et al. CORUM: the comprehensive resource of mammalian protein complexes — 2009. Nucleic Acids Res. 2010: p. D497–501.

[12] Koenigsknecht J, Landreth G. Microglial phagocytosis of fibrillar beta-amyloid through a B1 integrin-dependent mechanism. Neurobiol Dis. 2004: p. 9838-46.

[13] Ashburner M, Ball CA, Blake JA, Botstein D, Butler H, Cherry JM, et al. Gene Ontology: tool for the unification of biology. Nat Genet. 2000: p. 25-9.

[14] Hourani M, Mendes A, Berretta R, Moscato P. Genetic biomarkers for brain hemisphere differentiation in Parkinson's disease. In Computational Models for Life Sciences - CMLS '07. AIP Conference Proceedings; 2007. p. 207-16.

[15] Chung KKK, L. DV, Dawson TM. The role of the ubiquitin-proteasomal pathway in Parkinson's disease and other neurodegenerative disorders. Trends Neurosci. 2001: p. 7-14.

[16] Simunovic F, Yi M, Wang Y, Macey L, Brown LT, Krichevsky AM, et al. Gene expression profiling of substantia nigra dopamine neurons: further insights into Parkinson's disease pathology. Brain. 2008: p. 1795-809.

[17] Teismann P, Tieu K, Choi DK, Wu DC, Naini A, Hunot S, et al. Cyclooxygenase-2 is instrumental in Parkinson's disease neurodegeneration. Proc Natl Acad Sci U S A. 2003: p. 5473-8.

[18] Ma SL, Tang NLS, Zhang YP, Ji L, Tam CWC, Lui VWC, et al. Association of prostaglandin-endoperoxide synthase 2 (PTGS2) polymorphisms and Alzheimer's disease in Chinese. Neurobiol Aging. 2008: p. 856-60.

[19] Park L, Zhou P, Pitstick R, Capone C, Anrather J, Norris EH, et al. Nox2-derived radicals contribute to neurovascular and behavioral dysfunction in mice overexpressing the amyloid precursor protein. Proc Natl Acad Sci U S A. 2008: p. 1347-52.

[20] Imai H, Nakagawa Y. Biological significance of phospholipid hydroperoxide glutathione peroxidase (PHGPx, GPx4) in mammalian cells. Free Radic Biol Med. 2003: p. 145-69.

[21] Blackinton J, Kumaran R, van der Brug MP, Ahmad R, Olson L, Galter D, et al. Post-transcriptional regulation of mRNA associated with DJ-1 in sporadic Parkinson disease. Neurosci Lett. 2009: p. 8-11.

[22] Bonifati V, Rizzu P, van Baren MJ, Schaap O, Breedveld GJ, Krieger E, et al. Mutations in the DJ-1 gene associated with autosomal recessive early-onset parkinsonism. Science. 2003: p. 256–9.

[23] Chen L, Na R, Gu M, Richardson A, Ran Q. Lipid peroxidation up-regulates BACE1 expression in vivo: a possible early event of amyloidogenesis in Alzheimer's disease. J Neurochem. 2008: p. 197-207.

[24] Warner TT, Schapira AH. Genetic and environmental factors in the cause of Parkinson's disease. Ann Neurol. 2003: p. S16-23.

[25] Shi Z, Zhang J, Zheng S. What we know about ST13, a co-factor of heat shock protein, or a tumor suppressor? J Zhejiang Univ - Sci B. 2007: p. 170–6.

[26] Muchowski PJ, Wacker JL. Modulation of neurodegeneration by molecular chaperones. Nat Rev Neurosci. 2005: p. 11-22.

[27] Roodveldt C, Bertoncini CW, Andersson A, van der Goot AT, Hsu ST, Fernandez-Montesinos R, et al. Chaperone proteostasis in Parkinson's disease: stabilization of the Hsp70/alpha-sybuclein complex by Hip. EMBO J. 2009: p. 3758-70.

[28] Lancet D, Safran M, Olender T, Dalah I. GeneCards tools for combinatorial annotation and dissemination of human genome information. In GIACS Conference on Data in Complex Systems; 2008.

[29] Ji Y, Ferracci G, Warley A, Ward M, Leung KY, Samsuddin S, et al. beta-Actin regulates platelet nitric oxide synthase 3 activity through interaction with heat shock protein 90. Proc Natl Acad Sci U S A. 2007: p. 8839-44.

[30] Bassell GJ, Zhang H, Byrd AL, Femino AM, Singer RH, L. TK, et al. Sorting of beta-actin mRNA and protein to neurites and growth cones in culture. J Neurosci. 1998: p. 251-65.

[31] Ricciarelli R, d'Abramo C, Massone S, Marinari UM, Pronzato MA, Tabaton M. Microarray analysis in Alzheimer's disease and

normal aging. IUBMB Life. 2004: p. 349-54.

[32] Datta PK, Moses HL. STRAP and Smad7 synergize in the inhibition of transforming growth factor beta signaling. Mol and Cell Biol. 2000: p. 3157-67.

[33] Mogi M, Harada M, Kondo T, Narabayashi H, Riederer P, Nagatsu T. Transforming growth factor-beta1 levels are elevated in the striatum and in ventricular cerebrospinal fluid in Parkinson's disease. Neurosci Lett. 1995: p. 129-32.

[34] Vawter MP, Dillon-Carter O, Tourtellotte WW, Carvey P, Freed WJ. TGFβ1 and TGFβ2 concentrations are elevated in Parkinson's disease in ventricular cerebrospinal fluid. Exp Neurol. 1996: p. 313-22.

[35] 't Hoen PAC, Ariyurek Y, Thygesen HH, E Vreugdenhil E, Vossen RHAM, de Menezes RX, et al. Deep sequencing-based expression analysis shows major advances in robustness, resolution and inter-lab portability over five microarray platforms. Nucleic Acids Res. 2008: p. e141.

[36] Trimmer PA, Bennett Jr. JP. The cybrid model of sporadic Parkinson's disease. Exp Neurol. 2009: p. 320-5.

[37] Trimmer PA, Borland MK, Keeney PM, Bennett Jr JP, Parker WD. Parkinson's disease transgenic mitochondrial cybrids generate Lewy inclusion bodies. J Neurochem. 2004: p. 800-12.

[38] Esteves ARF, Domingues AF, Ferreira IL, Januário C, Swerdlow RH, Oliveira CR, et al. Mitochondrial function in Parkinson's disease cybrids containing an nt2 neuron-like nuclear background. Mitochondrion. 2008: p. 219-28.

[39] Barbanti P, Fabbrini G, Ricci A, Cerbo R, Bronzetti E, Caronti B, et al. Increased expression of dopamine receptors on lymphocytes in Parkinson's disease. Mov Disord. 1999: p. 764-71.

[40] Di Monte DA, Lavasani M, Manning-Bog AB. Environmental factors in Parkinson's disease. Neurotoxicology. 2002: p. 487-502.

[41] Hawkes CH, Del Tredici K, Braak H. Parkinson's disease: a dual-hit hypothesis. Neuropathol Appl Neurobiol. 2007: p. 599-614.

[42] Lerner A, Bagic A. Olfactory pathogenesis of idiopathic Parkinson disease revisited. Mov Disord. 2008: p. 1076-84.

[43] Phillips RJ, Walter GC, Wilder SL, Baronowsky EA, Powley TL. Alpha-synuclein-immunopositive myenteric neurons and vagal preganglionic terminals: autonomic pathway implicated in parkinson's disease? Neuroscience. 2008: p. 733-50.

[44] Halliwell B, Gutteridge JMC. Oxidative stress in PD. In Halliwell B, Gutteridge JMC, editors. Free Radic Biol Med; 1999; New York: Oxford University Press. p. 744-58.

[45] Whitton P. Inflammation as a causative factor in the aetiology of Parkinson's disease. Br J Pharmacol. 2007: p. 963-76.

[46] Su LJ, Auluck PK, Outeiro TF, Yeger-Lotem E, Kritzer JA, Tardiff DF, et al. Compounds from an unbiased chemical screen reverse both ER-to-Golgi trafficking defects and mitochondrial dysfunction in Parkinson's disease models. Dis Models Mech. 2010: p. 194-208.

[47] Li JY, Englund E, Holton JL, Soulet D, Hagell P, Lees AJ, et al. Lewy bodies in grafted neurons in subjects with Parkinson's disease suggest host-to-graft disease propagation. Nat Med. 2008: p. 501-3.

[48] Kordower JH, Chu Y, Hauser RA, Freeman TB, Olanow CW. Lewy body-like pathology in long-term embryonic nigral transplants in Parkinson's disease. Nat Med. 2008: p. 504-6.

[49] Shemin D, Rittenberg D. The life span of the human red blood cell. J Biol Chem. 1946: p. 627-36.

[50] Stuart M,J, Murphy S, A. OF. A simple nonradioisotope technic for the determination of platelet life-span. N Engl J Med. 1975: p. 1310-3.

[51] Muller-Sieburg C, Sieburg HB. Stem cell aging: survival of the laziest? Cell Cycle. 2008: p. 3798-804.

[52] Rossi D, Jamieson C, Weissman I. Stems Cells and the Pathways to Aging and Cancer. Cell. 2008: p. 681-96.

[53] Nijnik A, Woodbine L, Marchetti C, Dawson S, Lambe T, Liu C, et al. DNA repair is limiting for haematopoietic stem cells during ageing. Nature. 2007: p. 686-U9.

[54] Rafii S, Lyden D. Therapeutic stem and progenitor cell transplantation for organ vascularization and regeneration. Nat

Med. 2003: p. 702-12.

[55] Mouret A, Lepousez G, Gras J, Gabellec MM, Lledo PM. Turnover of newborn olfactory bulb neurons optimizes olfaction. J Neurosci. 2009: p. 12302-14.

[56] Natale G, Pasquali L, Ruggieri S, A. P, Fornai F. Parkinson's disease and the gut: a well known clinical association in need of an effective cure and explanation. Neurogastroenterol Motil. 2008: p. 741-9.

[57] Gewirtz AT. Intestinal epithelial pathobiology: past, present and future. Best Prac & Res Clin Gastr. 2002: p. 851-67.

[58] Creamer B, Shorter RG, Bamforth J. The turnover and shedding of epithelial cells. I. The turnover in the gastro-intestinal tract. Gut. 1961: p. 110-6.

[59] Paris F, Fuks Z, Kang A, Capodieci P, Juan G, Ehleiter D, et al. Endothelial Apoptosis as the Primary Lesion Initiating Intestinal Radiation Damage in Mice. Science. 2001: p. 293-7.

[60] Olanow CW, Prusiner SB. Is Parkinson's disease a prion disorder? Proc Natl Acad Sci U S A. 2009: p. 12571-2.

[61] Tamoa W, Imaizumia T, Tanjib K, Yoshida H. Expression of α-synuclein in vascular endothelial and smooth muscle cells. International Congress Series. 2003: p. 173-9.

[62] Zhang Y, Allodi S, Sandeman DC, Beltz BS. Adult neurogenesis in the crayfish brain: Proliferation, migration, and possible origin of precursor cells. Dev Neurobiol. 2009: p. 415-36.

[63] Asahara T, Toyoaki M, Sullivan A, Silver M, vanderZee R, Li T, et al. Isolation of Putative Progenitor Endothelial Cells for Angiogenesis. Science. 1997: p. 964-6.

[64] Barker N, van Es JH, Kuipers J, Kujala P, van den Born M, Cozijnsen M, et al. Identification of stem cells in small intestine and colon by marker gene Lgr5. Nature. 2007: p. 1003-8.

[65] Gonçalves JP, Grãos M, Valente AXCN. Polar Mapper: a computational tool for integrated visualization of protein interaction networks and mRNA expression data. J Royal Soc Interface. 2009: p. 881-96.

[66] J.-F. R, K. V, T. H, T. HK, Dricot A, Li N, et al. Towards a proteome-scale map of the human protein-protein interaction network. Nature. 2005: p. 1173-8.

[67] Stelzl U, Worm U, Lalowski M, Haenig C, Brembeck FH, Goehler H, et al. A human protein-protein interaction network: A resource for annotating the proteome. Cell. 2005: p. 957-68.

[68] Birnbaum ZW, Klose OM. Bounds for the variance of the Mann-Whitney statistic. Annals of Math Stat. 1957.

[69] Van Dantzig D. On the Consistency and Power of Wilcoxon's Two Sample Test. Koninklijke Nederlandse Akademie van Weterschappen, Series A. 1915.

[70] Barrett T, Troup DB, Wilhite SE, Ledoux P, Rudnev D, Evangelista C, et al. NCBI GEO: mining tens of millions of expression profiles — database and tools update. Nucleic Acids Res. 2007: p. D760–5.

5

Engineering in the Biological High-Throughput Regime

While Science aims to find predictive laws that are as broadly applicable as possible, Engineering endeavors to design and implement solutions for specific problems.

Besides opening new frontiers for Science, the recent technological advances in the Life Sciences are creating new opportunities for Engineering. In genetic engineering, fully de novo assembled microbes, could efficiently produce biofuel or stand ready to rapidly clean an environmental oil spill. In pharmaceutics, a cocktails of synergistic drugs, tailored to the detailed characteristics of both the pathology and the patient, could yield an increased therapeutic effect with nevertheless much reduced side-effects. In tissue engineering, expression of a key combination of factors in target cells could reprogram the fate of those cells to a desired state, thus regenerating a damaged tissue or correcting a pathological cell-state in a patient. Development of biological manipulation and assaying techniques will make feasible, or already makes feasible, implementation of the above engineering solutions. In Engineering, however, capacity to implement must be complemented by ability to design.

Traditionally, engineering relies on the principles of standardization and modularity to facilitate component-based design. These principles coupled with the relevant scientific knowledge allow rational formulation of a design. Although not perfect at first, the design is likely sufficiently close to functioning for a manageable amount of fine tuning, via iteration between implementation and redesign, to produce a satisfactory solution. This is the essence of the Classical Engineering Design (CED) framework that has been so successful across multiple engineering fields. We nevertheless argue that CED will be ineffective for realizing many of the engineering opportunities arising today in the Life Sciences, such as those in the examples given earlier. Predictive ability in the context of a living organism is too low for the CED approach to function. In particular, it is intrinsically difficult to anticipate interactions between different biological elements, which leads to a breakdown in the modularity pillar of CED. On the other hand, an approach based on blindly implementing and screening designs for a functioning one is futile, as even developments in high-throughput assaying cannot handle the exponentially large number of possible design-component combinations.

We propose jointly leveraging high-throughput assaying techniques and biological knowledge into a rationally optimized search of the space of design possibilities as the appropriate engineering philosophy to realize some of the opportunities arising in the Life Sciences today. We call this approach High-Throughput Biologically Optimized Search Engineering (HT-BOSE). Essentially, we advocate biological knowledge and insights are absolutely crucial for engineering design but, given the complexity of Biology, they are most fruitfully applied to optimizing the search process in the space of design possibilities. Generally, it is envisioned that there exist plenty of highly valuable synergies that can be exploited in a living system setting. We believe a HT-BOSE approach may provide the best chance of finding them. The article that follows discusses a HT-BOSE foundation for Synthetic Biology, the emerging field of engineered complex biological systems.

A. X. C. N. Valente

6

Article

High-Throughput Biologically Optimized Search Engineering Approach to Synthetic Biology

(arXiv: March 28th, 2011)

High-Throughput Biologically Optimized Search Engineering Approach to Synthetic Biology

A. X. C. N. Valente[1,2,3] and Stephen S. Fong[3,4]

[1]Biocant - Biotechnology Innovation Center, Cantanhede, Portugal
[2]Center for Neuroscience and Cell Biology, University of Coimbra, Portugal
[3]Center for the Study of Biological Complexity, Virginia Commonwealth University, Richmond VA, USA
[4]Dept. of Chemical and Life Sciences Engineering, Virginia Commonwealth University, Richmond VA, USA

Synthetic Biology is the new engineering-based approach to biology that includes applications of designing complex biological devices. At present, it is not yet clear what will emerge as the defining principles of Synthetic Biology. One proposed approach is to build Synthetic Biology around the classical engineering principles of standardization, modularity/decoupling and abstraction/modeling to facilitate component-based design. In this article we suggest and discuss an alternative paradigm, which we call High-throughput Biologically Optimized Search Engineering (HT-BOSE). Stemming from directed evolution, in HT-BOSE the focal point is a biological knowledge based rational optimization of the search process in the space of device design possibilities. The HT-BOSE approach may also be relevant in other contexts and we briefly highlight how it could be applicable to the development of multi-drug cocktails in a biomedical setting.

Introduction

Genetic engineering, defined as the purposeful manipulation of a single gene within an organism, is today a commonplace technology (1). A greater challenge is moving from single-gene manipulation, to the engineering of complex biological devices that incorporate multiple interacting functions. The goal at present is creation of a full-fledged engineering field with living matter as its substrate. This hopeful engineering field is most often called Synthetic Biology (2).

Every engineering field is characterized by its own set of canonical principles. These principles are distinct insofar as they are determined by different substrates of study, engineering objectives and relevant governing physical laws. Nonetheless, all major engineering fields are based on an overarching common design philosophy (2). Sustained by the three concepts of standardization of parts, modularity/decoupling and abstraction/modeling, this Classical Engineering Design (CED) approach allows the construction of complex engineering devices with manageable, limited trial and error tuning. This is remarkable, as a very large number of parts and an even larger number of interactions between those parts would normally render the successful implementation of a device all but impossible.

There is an ongoing major effort to build a CED foundation for Synthetic Biology (3). For instance, a registry of standard biological parts has been created (4). At present it contains thousands of distinct entries, with varying degrees of quality control for each part. This foundation already allows undergraduate students in a yearly summer Synthetic Biology competition to, in mere weeks, successfully build working devices of a few parts (5). However, the ultimate goal of Synthetic Biology is not being able to put together elementary devices in days or weeks, but rather to enable the engineering of complex devices.

In Synthetic Biology unwanted interactions are inherently difficult to predict and avoid (6). This is due to the molecular/chemical nature of the interactions and the unfeasibility of physically compartmentalizing every interaction. As the number of parts in a device grows, the number of such pernicious interactions will grow even faster. We argue this scaling poses a major hurdle for traditional CED component-based approaches and potentially renders CED an ineffective approach to engineer complex biological devices. For a device involving dozens of parts, the amount of fine tuning required would be unworkable.

A distinct conceptual view of engineering is attained by considering Design Space (DS) - the space of all possible device designs. Given the time to try out every possible design, it would be a mere formality to select the best design to the task at hand. Naturally, this is not feasible due to the very high dimensionality of DS. CED provides a way of immediately placing ourselves close to a working design, which can then be hopefully reached via some fine-tuning. However, as explained above, we argue that for complex biological devices, CED is not a good approach as in this case the large number of pernicious interactions will actually place us far from a functioning design.

The evolutionary histories of natural organisms trace paths in DS. This motivated the biological engineering technique of directed evolution (7). Directed evolution consists of performing alternating rounds of mutation and selection/screening with the intent of moving in DS towards a good design for the task at hand. It has been extensively applied to the engineering of enzymes (8), and more recently to the engineering of elementary biological devices (9). Nevertheless, so far it has not been possible to construct complex devices via directed evolution any more than it has been via CED. Directed evolution as a nonspecific statistical search technique in a very high-dimensional space of possibilities has not proven to be powerful enough.

We suggest that the future of Synthetic Biology lies in combining biological knowledge and directed evolution into a biologically-guided rational optimization of the DS search process. We designate this approach High-Throughput Biologically Optimized Search Engineering (HT-BOSE) and suggest it as a framework for designing biological devices. This article will focus on founding principles for HT-BOSE. In parallel, HT-BOSE will also require the development of appropriate technologies for its implementation, in particular for an increase in the throughput of the search process (10; 11; 12). We consider there are three foundational elements to an HT-BOSE approach, which we now discuss in turn: Search strategies, the Design Registry and fitness assays.

Search strategies

We start by posing a streamlined version of the search problem in HT-BOSE. Given a particular aim for a device, every design in DS has an associated fitness score. The fitness of any given design is unknown, until the design is realized and assayed in some fashion. The objective is to find a high fitness design, under the constraint that a limited number of designs can be assayed.

If there is no knowledge at all of the structure of the fitness function over DS, then the strategy of randomly picking designs in DS for assaying is as good as any other search strategy. This is an instance of the principle that "there are no free-lunches in optimization" (13).

Now, it is clear that designs that differ by a single base pair in their encoding DNA tend to have correlated fitness, as they tend to be designs with modest differences. This *DS functional structure* enables the pursuit of the traditional hill-climbing search, i.e. directed evolution, based on point mutations (7). Namely, a search that is based on successive rounds of point mutation and selection of the best mutated design(s) as the starting point(s) for the next round. Naturally, this hill climbing towards a higher fitness design is a much more efficient strategy than random search. Nonetheless, mutation interactions can be nonlinear with regards to effect on fitness. For instance, a simultaneous mutation in multiple sites might be beneficial, whereas the same mutations in isolation are not. As a result, many design improvements are unreachable via point mutation hill-climbing. In other words, point mutation hill-climbing is limited by local fitness optima.

Yet, there is a much richer DS functional structure than the one just described. For instance, there is DS functional structure at other length-scales. Knowledge of this additional biological structure can be explored to improve the hill-climbing search process. We start by describing optimization of the search process at a very different length-scale.

Microbial consortia

Community level synergies between different microbial species often result in a consortium of microbes being more efficient than a single organism at fulfilling a desired task (14). As an example, microbial communities might be pertinent to biofuel production (15). In the context of consortia, a point in DS represents a potential Community Design (defined by a design being ascribed to each organism in the community). A community can be optimized by the point mutation search process previously described, with the point mutations being applied over all the microbes in the community. However, there is also DS functional structure at the level of what microbes constitute the community. Therefore, an additional way of hill-climbing in DS is via whole-microbe-at-a-time changes to the community composition. Note how this constitutes searching DS via jumps at a distinct length-scale that is much larger than point mutation changes. In a sense, swapping microbes can be viewed as a course-grained adjustment within DS, whereas point mutations can be viewed as a fine-grained adjustment.

Given the biological significance of genes, there is invariably DS functional structure at the level of what genes are present in an organism. Hill-climbing in DS at the gene length-scale, via whole-gene-at-a-time changes to an organism, is therefore also pertinent. Gene length-scale changes would stand as an intermediate-grained adjustment compared to whole organism exchanges (course-grained) and genetic point mutations (fine-grained).

Recombination based techniques are widely used in the enzyme optimization context (16; 17). For this family of techniques, the hill-climbing in DS occurs via recombination of entire parts of proteins. The effectiveness of the method rests on the modular nature of proteins, that constitutes DS functional structure at yet another length-scale (18).

The examples above leverage generic DS functional structure. However, biological knowledge specific to the problem at hand can be invaluable. High-throughput genomic data collection and analysis can play an important role in this regard (19; 20; 21). As a case in point, many of these techniques yield functional networks of genes. Such networks embody DS functional structure information. Analogously to what happens with mutations, gene interactions can have a nonlinear effect on fitness. The simultaneous introduction of multiple genes in an organism might be beneficial, while the isolated introduction of any one of those same genes is not. Functional networks of genes enable hill-climbing strategies based on prioritizing specific multiple-genes-at-a-time changes to an organism. An otherwise unguided search at a multiple-genes-at-a-time-jumps length-scale would be generally unsuccessful, due to the volume of DS to be explored and the relative sparseness of worthy designs in DS. Functional networks at other biological levels may likewise be explored.

In CED, knowledge is often applied directly to the design of a device. It is not easy to acquire biological knowledge at the level of detail required for relevance in such a task. In HT-BOSE, biological knowledge is equally important. However, as shown, it is instead used to infer DS functional structure that enables a more efficient search process. We suggest this less demanding form of leveraging knowledge is more realistic for the complex biological setting and, in particular, may better match the typical detail-level of information generated by high-throughput approaches.

Ultimately, HT-BOSE will involve combining a diversity of hill-climbing strategies in a true multi-scale search process for engineering function. Figure 2 outlines how this could be implemented in practice. Now, a successful search in a very high-dimensional space, besides an efficient search strategy, requires an appropriate choice of where to start the search. We next discuss implications of this to an HT-BOSE approach to Synthetic Biology.

Design Registry for HT-BOSE
In CED, engineering of a new device relies on combining and reusing existing parts. A good library of standardized parts, appropriately catalogued and characterized, is crucial to enable this. For engineering of devices via a search process, the counterpart is a good library of *Seed Designs* to serve as search starting points. Further, such a Design Registry should also be a source of the seed-parts required for the ensuing jumps-at-different-length-scales search strategies discussed in the previous section.

A seed likely close in DS to a design that fulfills the engineering objectives at hand has obvious advantages. An existing microbe that already nearly fulfills the new engineering objectives constitutes naturally a good starting point for modifications. However, the choice of ideal starting point is not always so clear. There may be no existing device already very close to fulfilling the desired engineering objectives. For example, the objectives may require a combination of traits of vastly different existing organisms. An important characteristic for a seed is that of flexibility. Some designs may not be practical engineering devices for any purpose and yet constitute excellent seeds. That is the case if the structure of DS is such that it is possible to hill-climb from them to a large diversity of useful designs. Again, from the versatile-community to the versatile-single gene, such flexible designs and design-parts are present at all levels.

In contrast to the versatile design, a design may be an excellent practical device for a given purpose and yet, live in a region of DS that makes it a not so good starting point for hill-climbing to a design that addresses any objective other than that one. A trade-off between specificity (in the sense of closeness to fulfilling the final desired objective) and versatility may have to be considered when choosing seeds for a search process. The successful selection of seeds will not be trivial and likely will require trial and error. It will constitute a major part of the process of engineering a new device via HT-BOSE.

The seed selection process will rest on information available in the Design Registry. A design entry in the Registry must, besides data on functional characteristics, record how that design was generated. When an existing design is used as a seed for subsequent designs, this must also be noted. Effectively, designs can have a forward phylogenetic history in addition to their backward phylogenetic history. Sliding along a phylogeny may allow in particular an adjustment of the specificity-flexibility trade-off in seed selection.

The complexity of the phylogenetic history of a design is compounded by the fact that the search strategies discussed do not necessarily lead to elementary relationships between designs. The search at different length-scales process creates devices that have received contributions from a multiplicity of sources, across a range of length-scales. The discovery of horizontal gene transfer has complicated natural phylogenetics, but a phylogenetics for designs resulting from HT-BOSE is likely to be even more complex. Developing an ontology that facilitates handling of all this information will be of essence, given the crucial role of seed selection in the HT-BOSE engineering approach.

In its early stages, a Design Registry for HT-BOSE should take maximum advantage of the enormous diversity of communities, microbes and genes that Nature already provides. In this regard, it stands to benefit greatly from large-scale metagenomic data collection efforts (22; 23). As new synthetic designs are engineered and added to the Design Registry, older entries gradually develop forward phylogenies. In due course, the more flexible designs and design-parts become apparent, making the Design Registry ever more valuable in facilitating the seed selection phase of HT-BOSE.

We have discussed seed designs, where DS search processes begin. We have also already addressed search strategies for generating designs to assay in subsequent rounds of a search process. It remains to focus on the third major element of an HT-BOSE approach to Synthetic Biology, the fitness scoring assay.

Fitness assays

Fitness assaying is the process whereby select designs are assayed and assigned a fitness score. For HT-BOSE, the assaying and fitness scoring should ideally be amenable to high-throughput implementation. The best designs, based on their scores, are then utilized by the search strategy to generate the next set of designs to fitness assay. Iteration of this procedure hopefully leads to hill-climbing in DS towards a design that accomplishes the engineering objective at hand.

The natural fitness assay to use is the one that assesses how well a design fulfills the desired engineering objective. Unfortunately, this straightforward assay choice may not necessarily succeed. When the designs being assayed are still far from accomplishing the desired engineering purpose, a fitness score based on proximity to satisfying that goal may lack the *scoring resolution* for differentiating between those designs. The designs may all appear equally bad at achieving the objective, therefore precluding hill-climbing via prioritization of the best ones. In such situations, the HT-BOSE approach requires devising a succession of *transitional fitness assays*. These transitional fitness assays must provide intermediate goals that are interspersed closely enough for adequate fitness scoring resolution to exist at all times. Biological insight must be used to ensure that fulfillment of the succeeding intermediate goals corresponds to getting progressively closer to a design that meets the final engineering goal. Successful transitional fitness assays are likely rather problem-specific and therefore their formulation constitutes one of the more challenging tasks in HT-BOSE.

A switch to hill-climbing under an alternative *supporting fitness assay* may be valuable in instances the search in DS gets trapped in a local optima with respect to the main (and ultimately relevant) fitness function. Again, the choice of a supporting fitness assay should rely on

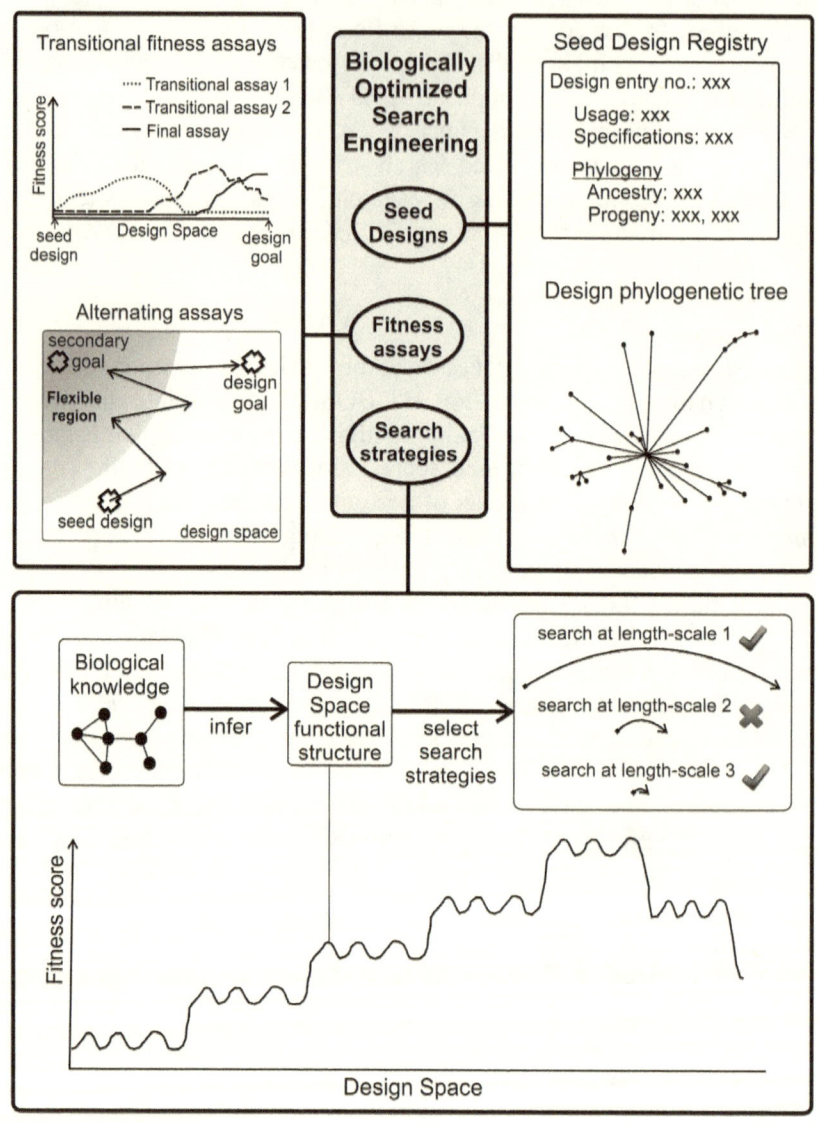

Figure 1 (Opposite page). The HT-Biologically Optimized Search Engineering approach to Synthetic Biology.
Seed Designs: A Registry of Seed Designs provides starting points (seeds) for the search processes in Design Space. To facilitate seed selection, Registry records should include phylogenetic information on how designs were generated, as well as on what subsequent designs originated from them. Seed selection may involve considering the trade-off between closeness of a seed to fulfilling the desired engineering objective and versatility of the seed as a search starting point. Designs that show high versatility as search starting points are apparent in the design phylogenetic tree. **Fitness assays:** Fitness scores are assigned to designs tested in the fitness assay. If assayed designs are still too far from fulfilling the desired goal, an evaluation based on this fulfillment may be unable to differentiate between the designs. In such cases, a succession of transitional fitness assays associated with intermediate engineering goals must be employed. Search processes can also get trapped in local fitness optima. Alternating the main assay with a supporting assay may circumvent this problem. A supporting assay that draws the search towards a more flexible region of Design Space may prove particularly helpful. **Search strategies:** Biological data is seldom detailed enough to allow a direct design approach. Yet, it may permit inference of Design Space functional structure. Devising rational Design Space search strategies is then possible. In particular, Design Space functional structure at multiple length-scales should be explored.

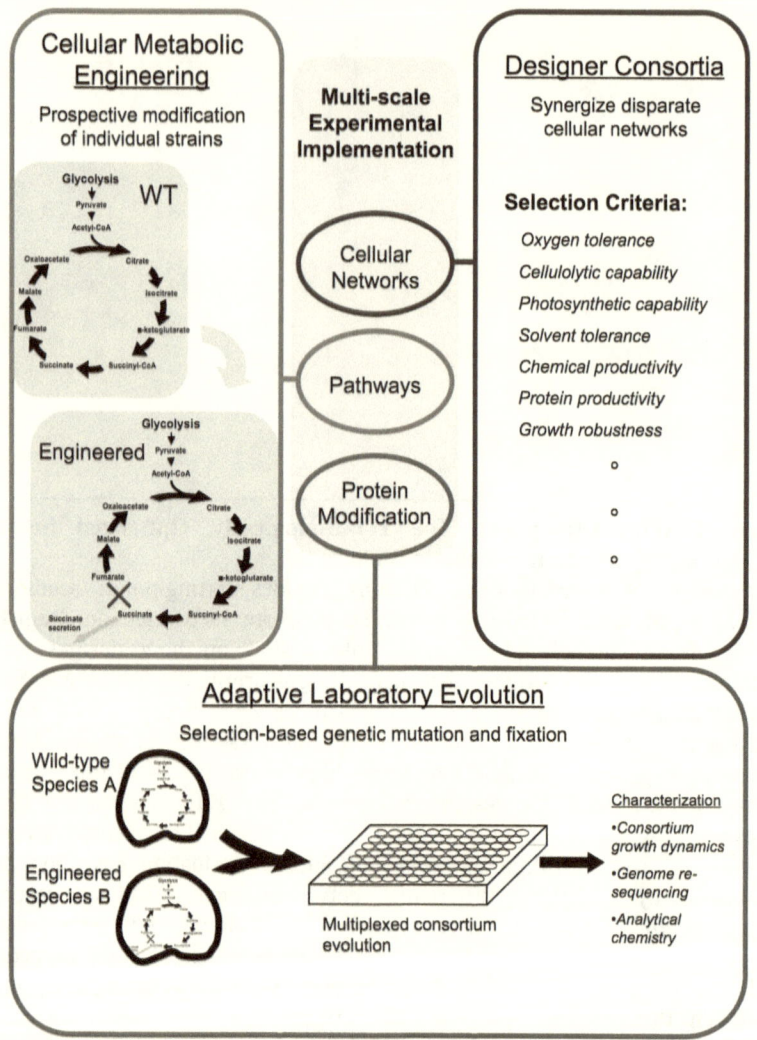

Figure 2. Outline of a microbial consortia optimization experiment.
Multi-scale experimental implementation of HT-BOSE will utilize cellular, pathway and molecular/genetic experimental tools to parallel computational analyses. Different organisms can be screened and selected based upon favorable functional characteristics to define the initial composition of a **Designer consortium**. Each of the selected organisms can be modified by pathway engineering techniques to delete or add biochemical capabilities (**Metabolic engineering**). Molecular and genetic changes can be implemented through extended co-culturing of selected and/or engineered organisms to foster stable consortia using **Adaptive laboratory evolution**.

biological insight into the specific problem in question. An intelligent choice for the supporting fitness assay could greatly surpass the typical default option of random jumping in DS every time trapping in a local optima occurs. A particularly relevant characteristic for a supporting assay may be the ability to draw the search towards a more versatile region in DS, before hill-climbing of the main fitness function towards the ultimate goal is resumed. Note how the specificity-flexibility adjustment, previously mentioned in the context of seed design selection, is now occurring during the actual search, via alternating between the main and the supporting fitness assay. A case in point is the use of a thermostability assay as the supporting assay in enzyme optimization problems (24).

Dynamically, optimization under a single fitness function can be visualized as a straightforward monotonic hill-climb towards a local fitness peak. By comparison, optimization under alternating fitness functions can produce a much richer spectrum of dynamical behaviors. As the simplest example, periodic orbits are possible in this latter case. It will be interesting to further explore from a mathematical standpoint this class of switched dynamical systems.

We end this section with reminders related to two peculiarities of engineering in a biological context. First, synthetic biological devices must be engineered for evolutionary stability under the setting and time-scale on which they will function. Therefore the fitness assay must be formulated to properly assess the operationally relevant evolutionary end point of a design. Second, a fundamental premise for pursuing HT-BOSE is that unwanted biological interactions are inherently difficult to circumvent. Therefore the subject of any fitness assaying should be the full integrated device at all times. Unlike under a CED approach, it makes no sense to develop different parts of a device in isolation and combine them only at a later stage.

Although we have been discussing HT-BOSE solely in the context of Synthetic Biology, it is applicable to other problems involving search within a very large space of possibilities. In the next section, we highlight the relevance of HT-BOSE in a particular biomedical setting.

HT-BOSE in medical drug discovery

Medical drug discovery has been dominated by the single compound, single molecular target, paradigm. However, many common diseases, such as cancer, diabetes, immune-inflammatory and cardio-vascular disorders are likely inherently system-level pathologies (25). There is a growing appreciation that trying to address such system-level disruptions via modulation of a single target is unnecessarily restrictive.

Instead, the single drug magic-bullet could be beneficially replaced by the multi-compound drug cocktail, that synergistically acts at multiple points of the system (26; 27). But direct design of a drug cocktail is too difficult while, akin to the DS of Synthetic Biology, the Cocktail Space of a priori possible drug cocktails is too large for a blind exhaustive search. A credible alternative may be biologically rational optimization of the search process. The HT-BOSE approach thus becomes directly relevant to the drug cocktail discovery problem. As an elementary example, Drug Cocktail Space functional structure exists in the form of compounds that are co-present in natural organisms. Medically this is reflected in the oftentimes apparent more positive therapeutic effect of a traditional medicine natural product versus that of its purified, most significant compound in isolation (27).

Final Remarks

We propose that applying biological knowledge to the rational optimization of a DS search process is the most appropriate paradigm for Synthetic Biology. We discussed an elementary foundation for this HT-BOSE approach. In particular, we indicated how in HT-BOSE biological knowledge would be leveraged: via appropriate choice of seed designs, search strategies (based on DS functional structure) and transitional and supporting fitness assays. Notably, the HT-BOSE approach underscores harnessing the unforeseeable, but likely plentiful and highly valuable, synergies possible between existing biological elements.

Although we stressed contrasts between the two, there is effectively a continuum from pure HT-BOSE to pure CED. Different Synthetic Biology problems will likely require approaches at different points within this spectrum, seldom at the two extremes.

Synthetic Biology and Systems Biology, if successful, will bring with them a greatly enhanced understanding of living organisms, through the ability to engineer them and the ability to predict their behavior (28; 29). Hopefully the power of this knowledge will lead to an ever greater respect and appreciation for the preciousness of Life (30; 31).

References

1. Nicholl, Desmond S. T. *An Introduction to Genetic Engineering.* s.l. : Cambridge University Press, 2008.

2. *Foundations for engineering biology.* Endy, Drew. 2005, Nature, Vol. 438 (24), pp. 449-453.

3. *Refinement and standardization of synthetic biological parts and devices.* Barry Canton, Anna Labno, Drew Endy. 2008, Nature Biotechnology, Vol. 26 (7), pp. 787-793.

4. Registry of Biological Parts. [Online] http://partsregistry.org.

5. *iGEM.* [Online] http://ung.igem.org.

6. Ernesto Andrianantoandro, Subhayu Basu, David K. Karig, Ron Weiss. *Synthetic biology: new engineering rules for an emerging discipline.* 2006. Vol. 2, article number: 2006.0028.

7. *Design by directed evolution.* Arnold, Frances H. 1998, Accounts of Chemical Research, Vol. 31 (3), pp. 125–131.

8. Frances H. Arnold, George Georgiou (editors). *Directed Enzyme Evolution: Screening and Selection Methods (Methods in Molecular Biology).* s.l. : Humana Press, 2010.

9. *Directed evolution of a genetic circuit.* Yohei Yokobayashi, Ron Weiss, Frances H. Arnold. 2002, Vol. 99 (26), pp. 16587-16591.

10. *Diversity-based, model-guided construction of synthetic gene networks with predicted functions.* Ellis, Tom, Wang, Xiao and Collins, James J. 2009, Nature Biotechnology, Vol. 27, pp. 465 - 471 .

11. *Combinatorial synthesis of genetic networks.* Guet, Calin C., et al. 2002, Science, Vol. 296, pp. 1466-1470.

12. *High-throughput metabolic engineering: Advances in small-molecule screening and selection.* Dietrich, Jeffrey A., McKee,

Adrienne E. and Keasling, Jay D. 2010, Annu. Rev. Biochem., Vol. 79, pp. 563–90.

13. *No free lunch theorems for optimization.* D. H. Wolpert, W. G. Macready. 1997, IEEE Transactions on Evolutionary Computation , Vol. 1 (1), pp. 67 - 82.

14. Brenner, Katie, You, Lingchong and Arnold, Frances H. *Engineering microbial consortia: a new frontier in synthetic biology.* 2008. pp. 483-489 . Vol. 26 (9).

15. Eiteman, Mark A., Lee, Sarah A and Altman, Elliot. *A co-fermentation strategy to consume sugar mixtures effectively.* 2008. Vol. 2 (3).

16. *When blind is better: Protein design by evolution.* Arnold, Frances H. 1998, Nature Biotechnology, Vol. 16, pp. 617-618.

17. *Fancy footwork in the sequence space shuffle.* Arnold, Frances H. 2006, Nature Biotechnology, Vol. 24 (3), pp. 328-330.

18. *On the conservative nature of intragenic recombination.* Drummond, D. Allan, et al. 2005, Proceedings of the National Academy of Sciences, U.S.A., Vol. 102 (15), pp. 5380-5385.

19. *Systematic mapping of genetic interaction networks.* Dixon, Scott J., et al. 2009, Annual Review of Genetics, Vol. 43, pp. 601-625 .

20. *Functional organization of the yeast proteome by a yeast interactome map.* André X. C. N. Valente, Seth B. Roberts, Gregory A. Buck, Yuan Gao. 2009, Proceedings of the National Academy of Sciences, U.S.A., Vol. 106 (5), pp. 1490-1495 .

21. *Genetic reconstruction of a functional transcriptional regulatory network.* Hu, Zhanzhi, Killion, Patrick J. and Iyer, Vishwanath R. 2007, Nature Genetics, Vol. 39, pp. 683 - 687 .

22. *Metagenomics and industrial applications.* Lorenz, Patrick and Eck, Jürgen. 2005, Nature Reviews Microbiology, Vol. 3, pp. 510-516 .

23. *The Sorcerer II Global Ocean Sampling Expedition: Northwest Atlantic through Eastern Tropical Pacific.* Rusch1, Douglas B., et al. 2007, PLoS Biology, Vol. 5 (3), p. e77.

24. *In the light of directed evolution: Pathways of adaptive protein evolution.* Bloom, Jesse D. and Arnold, Frances H. 2009, Proceedings of the National Academy of Sciences, USA, Vol. 106, pp. 9995–10000.

25. *Genomics and medicine: an anticipation. From Boolean Mendelian genetics to multifactorial molecular medicine.* 2000, CR Acad. Sci. III, Vol. 323, pp. 1167-1174.

26. *Multi-target therapeutics: when the whole is greater than the sum of the parts.* Zimmermann, Grant R., Lehár, Joseph and Keith, Curtis T. 2007, Drg Discovery Today, Vol. 12, pp. 34-42.

27. *Multicomponent therapeutics for networked systems.* Keith, Curtis T., Borisy, Alexis A. and Stockwell, Brent R. 2005, Nature Reviews Drug Discovery, Vol. 4, pp. 71-78.

28. *Prediction in the hypothesis-rich regime.* Valente, André X. C. N. 2010, arXiv, p. 1003.3551v1.

29. *Systems biology and new technologies enable predictive and preventative medicine.* Hood, Leroy, et al. 2004, Science, Vol. 306, pp. 640-643.

30. *Synthetic biology and the ethics of knowledge.* Douglas, Thomas and Savulescu, Julian. 2010, Journal of Medical Ethics, Vol. 36, pp. 687-693.

31. *The promise and perils of synthetic biology.* Tucker, Jonathan B. and Zilinskas, Raymond A. 2006, The New Atlantis, Vol. Spring, pp. 25-45.

www.ingramcontent.com/pod-product-compliance
Lightning Source LLC
Chambersburg PA
CBHW022118170526
45157CB00004B/1682